The Ecological Challenge

Ethical, Liturgical, and Spiritual Responses

Dianne Bergant, C.S.A.
Edward Foley, Capuchin
Richard N. Fragomeni
Kathleen Hughes, R.S.C.J.
John Manuel Lozano, C.M.F.
Thomas A. McGonigle, O.P.
Thomas A. Nairn, O.F.M.
Gilbert Ostdiek, O.F.M.
John T. Pawlikowski, O.S.M.
Rabbi Hayim G. Perelmuter
Barbara Reid, O.P.
Paul J. Wadell, C.P.

Richard N. Fragomeni
and
John T. Pawlikowski, O.S.M.
Editors

A Michael Glazier Book
THE LITURGICAL PRESS
Collegeville, Minnesota

A Michael Glazier Book published by The Liturgical Press

"The Preparatory Rites: A Case Study in Liturgical Ecology" by Edward Foley, Kathleen Hughes, and Gilbert Ostdiek originally appeared in *Worship* 67:1 (January 1993) 17–37. It is reprinted with the permission of the editor.

Cover design by Mary Jo Pauly.

1 2 3 4 5 6 7 8

Library of Congress Cataloging-in-Publication Data

The ecological challenge : ethical, liturgical, and spiritual responses / Richard N. Fragomeni and John T. Pawlikowski, editors.
p. cm.
"A Michael Glazier book."
Includes bibliographical references.
ISBN 0-8146-5840-7
1. Human ecology—Religious aspects—Christianity. I. Fragomeni, Richard N. II. Pawlikowski, John.
BT695.5.E28 1994
261.8'362—dc20 93-42758
CIP

Contents

Introduction

Ecological awareness has been a growing phenomenon in Western society for well over two decades. But it has been somewhat slow in taking root in Catholic consciousness. Part of this no doubt has been the result of explicit hostility to Christianity on the part of some of the founders of the contemporary ecological movement, who directly implicated the biblical tradition in the rise of ecological deterioration. The symbol for Earth Day, of pagan origins, was deliberately selected to make this point. Some Christian leaders who have memories of the Nazi era have also expressed fears that the ecological movement may be leading us back to a glorification of naturalism which proved so destructive during that period of mass annihilation.

But it is time for the Christian community to cast aside these fears. They pale in significance when we face squarely the reality that the social philosopher Hans Jonas has placed before us, namely, that we are the first generation that must decide whether or not our progeny will inherit a world capable of sustaining life. In the past, anything destructive humankind could do to nature could be healed through nature's inherent recuperative powers. But increasingly that no longer seems to be the case. The rain forests will not return automatically if we destroy them. The ozone layer that protects us from the sun's harmful rays may not be restored if we allow its further erosion. Species vital to our food chain may not reappear if we do not take immediate steps to stop their rapid disappearance.

Fortunately, some sectors of Catholicism are beginning to take notice of the gravity of the ecological challenge. Pope John Paul II's 1990 World Peace Day address on this topic provided an important stimulus. And he emphasized the importance of ecological responsibility in his 1991 encyclical *Centesimus Annus.*

Episcopal conferences in the Philippines, Poland, and the Dominican Republic have boldly addressed this issue in recent years. Here in the United States Bishop Anthony Pilla of Cleveland and the bishops of Florida were the first Catholics on the leadership level to call for a new sense of ecological commitment within the Catholic community. At their November 1991 semiannual meeting, the full U.S. Conference of Bishops gave final approval to a major statement on the Catholic response to the new ecological challenge (see "Renewing the Earth," *Origins* 21:27, December 12, 1991). And Cardinal Joseph L. Bernardin of Chicago, in a plenary address to the fourteenth meeting of the International Catholic-Jewish Liasion Committee in Baltimore in May 1992, spoke of the need for human recovery of a sense of dependence upon the world as critical for the survival of humankind and the planet as a whole (see "*Tikkun Olam:* Healing the World," *Catholic International* 3:13 [1–14 July 1992]).

If this growing awareness of the need for ecological preservation within Catholicism, and within Christianity generally, is to become a permanent feature of the Church's consciousness, it will need to be undergirded by sound ethical reflection and the development of a spirituality and of worship patterns in which this awareness becomes a central motif. The essays in this volume try to make a modest beginning towards attainment of that goal. Many of them were first developed as part of a faculty seminar at the Catholic Theological Union during the academic year 1990–91.

In part 1, on Scripture, Dianne Bergant addresses the issue of how the Hebrew Scriptures, so often criticized as a fundamental source of ecological irresponsibility, view the relationship between humanity and the rest of creation. From a careful analysis of Hosea 2 she concludes that at least for an important strain of thought within the Hebrew Scriptures God is seen as establishing a relationship with the world which is integrally related to the structures of the natural world. Clearly this line of thinking rules out any form of anthropocentric imperialism. Dianne Bergant clearly establishes the continuing value of the biblical tradition in the development of a spirituality for contemporary ecological commitment, contrary to what some in the Earth Day movement have claimed.

Barbara Reid examines the writings of St. Paul in light of the contention by "ecologian" Thomas Berry that the biblical story of creation needs radical revision in the Ecozoic age which humankind is now entering. She does not necessarily disagree with Berry's call for a radical revision. Her argument is rather that there is a basis for such a revision in Pauline thought, particularly in Paul's teachings about the new age of salvation as "new creation" in Galatians and 2 Corinthians and in his reflections on the "body of Christ" in Romans and I Corinthians. She also feels that Paul's perspectives on the need for self-sacrifice (rooted in Christ's own sacrifice on the cross) can help deepen Berry's insistence on sacrifice as an integral part of survival in the Ecozoic age. While Pauline thought may be limited in its ability to provide a spiritual context for ecological commitment, Professor Reid is convinced that it contains valuable resources which should not be overlooked in present-day discussions.

The three essays in part 2, on ethics, try to show where the Catholic tradition (at least on the official level) has stood on the matter of ecology

(Thomas Nairn), how important theological themes such as the incarnation and eschatology relate to the development of a sense of the ethical in the realm of ecology (John Pawlikowski), and what ought to be our basic model for understanding the relationship between humanity and the rest of creation if we wish to develop a sound ecological ethic for our time (Paul Wadell). Each of the contributors to this section show that possibilities for a sound eco-theology exist within the Christian biblical, magisterial, and theological traditions. But they also stress the need to shake loose from certain perspectives, beliefs, and frameworks, some of which have been part of the Christian tradition for centuries.

Part 3 takes up the question of liturgy and ecological responsibility. This is a vital dimension of the current ecological discussion that has received too little attention up till now. Recognizing that human beings are a blend of rationality and vitality and that ethical response depends as much on the latter as the former, we need to come to a better appreciation of the indispensable role of liturgical celebration in the formation of ethical awareness and response. The various contributors to this section are all keenly aware of this reality. Professor Richard N. Fragomeni focuses on the writings of Walter J. Ong, S.J., Thomas Berry, C.P., and Brian Swimme (a mathematical cosmologist) in discussing the regretable loss of creational elements in recent liturgical reforms and the failed effort to restore a creational awareness in the proposed 1980 "Eucharistic Prayer A," which never received the approbation of the U.S. hierarchy or Vatican authorities. He suggests ways in which the insights of Ong, Berry, and Swimme can enhance creational consciousness even within the setting of present liturgical texts and formats (e.g., the paschal vigil).

In a joint effort Professors Edward Foley, Kathleen Hughes, and Gilbert Ostdiek offer a detailed analysis of the liturgical rites of Eucharistic preparation. Much in the same vein as Fragomeni, they argue that the creational dimensions have been pushed to the periphery or left completely aside and suggest ways in which this might be corrected, so that participation in the Eucharist might convey the sense of creational responsibility as an integral part of the sacrament.

Part 4, on spirituality, brings us face-to-face with the tradition as well as with important trends in U.S. Catholicism. Professor Thomas McGonigle, examining the thought of Hugh of St. Victor, shows that there existed a positive appreciation of the natural world in medieval spirituality which has often been overlooked. Without overstating their significance, important medieval texts can prove of significant value in the current search for an ecologically sound Christian faith perspective. John Manuel Lozano moves us into the American context, examining some of the writings of celebrated authors on spirituality such as Thomas Merton, as well as somewhat lesser-known travel accounts of early religious order members who journeyed throughout the continent to establish new foundations for their communities. Both of these bodies of literature show a remarkable appreciation for human interaction with the natural world in the emergence of genuine spirituality.

In our final contribution, Dr. Hayim Perelmuter devotes attention to two significant facets of the Jewish tradition that have important implications for the ecological discussion. In the first part of his essay he shows how the original scriptural text is not necessarily the final "word of God" for the Jewish tradition. This becomes especially relevant for the much-discussed (and often criticized) texts in Genesis which some claim are the basis for human attitudes of disregard for the rest of creation. Rabbi Perelmuter shows how the authentic meaning of these texts from the Jewish perspective is to be found only in the later commentaries, where they clearly are interpreted in a way that favors ecological sensitivity rather than insensitivity.

His second point is also crucial. He surveys the various liturgical festivals of Judaism to show how they enhance awareness of creational linkage. It is this linkage that has been underplayed, or even completely lost, in Christian liturgical celebration, as we see in chapters 6 and 7. Studying Jewish liturgy, in which the Christian liturgical tradition has deep roots, will further the process of retrieving creational responsibility in Christian liturgy advocated by Fragomeni, Ostdiek, Hughes, and Foley.

Professor Perelmuter's contribution clearly demonstrates how important it is for the contemporary Church to recapture its Jewish roots. Ecology is one area where the Christian tradition suffered impoverishment from what biblical scholar Cardinal Carlo Martini of Milan has called the Church's first and most devastating schism, its rupture with the Jewish people and the covenantal traditions of Israel.

The ten essays in this volume certainly do not presume to provide all that is needed for the emergence of a deep and abiding sense of creational responsibility in the Christian faith community. Hopefully, however, they make a good start towards that crucial goal. Many important questions remain: How should we interpret the varied texts of Scripture and what criteria do we use in selecting certain ones for use today? How far should we think of humanity in terms of a creational whole instead of the hierarchical model to which we have been long accustomed? Is the "organic model" the right road to take? Should we merge the human species into the larger context of creation, as in much of Eastern art, or still maintain some type of pyramid with the human community retaining a definite directional role over the rest of creation? How can we make creational responsibility part of the core expression of our liturgical celebration without falling into the trap of creating "theme" liturgies (a project which most liturgists severely criticize)? How far do we wish to go in understanding the natural world as a source of divine presence and energy without falling into a Nazi-like glorification of nature? And to what extent is the contemporary development of ecological foundations for spirituality and ethics a matter of retrieval and to what extent does it involve the search for totally new "revelation" (as Thomas Berry would argue)?

The basic conviction behind this volume is that a spirituality generated ethical commitment to ecological preservation is the greatest challenge now facing the Christian community. Ecology is not a matter of technological know-how or financial resources alone, though both remain important. Ultimately it requires a new (or in part renewed) faith perspective in which humanity enters a relationship with the rest of creation involving a blend of partnership as well as leadership of unprecedented proportions. In short, it requires a new ethic, a new spirituality, and a new liturgy.

Richard N. Fragomeni
John T. Pawlikowski, O.S.M.

Part I

Scriptural Perspectives

1

Restoration as Re-creation in Hosea 2

Dianne Bergant, C.S.A.

Peace With All Creation

For too long some people have viewed ecologists as single-issue advocates, specialists more concerned with the environment than with people. These same people have regarded immediate human need and legitimate ambition as far more important and pressing than environmental conservation and ecological restoration. Perhaps in general the gravity of the ecological issue has not been realized because women and men have not been attentive to the limits of the natural wealth of the world, for to date, the world has been prodigal in surrendering its treasures to those who know how to exploit them. Or, if some people have been conscious of nature's limits, they may have disregarded them, believing that the wealth of the world was theirs for the taking.

The environmental catastrophes of the 1980s have brought the seriousness of this issue to the forefront. Oil and chemical spills, acid rain, global warming and the greenhouse effect, the destruction of rain forests, droughts, and the nuclear accident at Chernobyl, with their tragic losses or crippling of lives, are only a few incidents we have seen that seem to threaten the ecological balance of the entire planet. What was previously judged to be an esoteric interest of marginal individuals and groups is now recognized as a central concern of people around the globe. Everywhere efforts are being made to bring ecological questions to the forefront of world consciousness and to make the environment a major public issue.

In 1967, Lynn White, Jr., a cultural historian, blamed Christianity for what has become the environmental crisis. He claimed that the Bible granted a "dominion mandate" which resulted in a kind of human imperialism that ruled at the expense of the rest of the natural world. Although his accusation has been challenged by various Christian theologians and ecologists, we cannot deny having been taught that human beings were told by God to conquer or subdue the earth and have dominion over the fish and the birds and every living thing (cf. Gen 1:26, 28). This kind of teaching was further illustrated in some catechisms by a pyramid divided, from base to top, into mineral creation, vegetation, animate creation, and humankind. Denying White's charge will neither exonerate the Church nor remedy the strain on the environment. Instead, our attention must be given to the ecological problem and to the theological issues that surround it.

The bishops of the United States, through the publication of their pastorals on peace and on the economy, have brought these two serious problems to national consciousness. In both documents, concern for the natural world is clearly stated. In *The Challenge Of Peace* we read: ". . . true peace implied a restoration of the right order not just among peoples, but within all of creation" (no. 32). Similarly, *Economic Justice For All* states: "To stand before God as the Creator is to respect God's creation, both the world of nature and of human history" (no. 34). More recently they have published *Renewing the Earth,* a reflection devoted to environmental issues, as well as *The Harvest of Justice Is Sown in Peace,* which also addresses this pressing question. Despite such statements from church leaders, many people committed to the directions set by these two pastorals still overlook the importance of ecological concerns in their commitment to peace in the world and economic justice. Such disregard can no longer continue.

Church officials and theologians have been accused of being late in addressing environmental issues. While this may well be true of some, many people are indeed confronting the problem. Recently, several episcopal conferences (e.g., Dominican Republic, 1987; the Philippines, 1988; Indonesia, 1989; Lombardy, 1988) have issued pastoral letters addressing the crisis. The universality of the situation was recognized by Pope John Paul II, who devoted his 1991 World Day of Peace message, entitled "Peace With God the Creator, Peace With All Creation," to this question. There he insists that the ecological crisis is a serious moral crisis and he exhorts women and men everywhere, whether or not they are motivated by religious conviction, to assume responsibility for our present situation and to join together in an effort to restore the health of the environment. The urgency of this problem has also been addressed by individual bishops (e.g., Anthony Pilla of Cleveland, 1990, and Michael Pfeifer of San Angelo, Texas, 1990).

Theologians too have committed themselves to the development of principles of eco-justice. The focus and breadth of this term are explained by one writer as follows:

> A combination of ecology and social justice, "eco-justice" refers to the interlocking web of concern about the earth's carrying capacity, its ability

> to support the lives of its inhabitants and the human family's ability to live together in harmony. It highlights the interrelatedness of such pressing issues as world hunger and world peace, the energy crisis and unemployment, appropriate technology and good work, biblical stewardship and feminist consciousness, racial justice and pluralistic community, life-style choices in response to poverty and pollution. The historic questions wherein God speaks with loving justice are whether humans will permit each other to live and enjoy a fair share of the fruits of the earth, and whether human activity will respect the limits of the earth's ability to support the community of life into the future (Hessel, 1985, 12).

Recognizing that a new socio-historical context often calls for a new reading of theology, some theologians have undertaken a second task that may be even more difficult than the development of eco-justice principles. They are engaged in the reinterpretation of those specific theological perspectives that may have contributed to a world view detrimental to ecological balance. This eco-justice concern has prompted biblical theologians to reexamine the use of nature imagery in the biblical testimonies to God's self-revelation. Meanwhile, many have come to wonder whether perhaps our familiarity with certain passages and our own anthropocentric world view have limited our reading of some of the Bible.

Creation and Convenant

Nature and nature imagery play a prominent role in the Book of Hosea. In fact, nature itself is the critical issue facing the prophet. What or who is the source of fertility and how is fruitfulness maintained? In an attempt to ensure fecundity, the people had turned away from YHWH to the Baals. In formal prophetic speech, Hosea condemned this breach of faith, threatened barrenness and desolation as punishment and then, in a surprising reversal, promised a new creation teeming with life and productivity. In the lyrical style of a poet, he used nature imagery to describe both the defilement and disintegration of the covenant between YHWH and the nation and, as well, its eventual reestablishment.

Issues of eco-justice have prompted biblical interpreters to ask new questions of Hosea's use of nature imagery. Was it primarily a kind of figurative speech or did it flow from a noteworthy mythic world view that perceived the interrelationship of all creation in a way that a scientific world view does not? Does Hosea view the natural world as merely serving the self-interests of women and men, even if these self-interests conform to the conditions of the covenant, or does he revere nature as having intrinsic value independent of humankind?

Since reading can never be detached from the socio-historical context of the reader, questions such as these constitute the lens through which I will examine a key passage in the second chapter of the Book of Hosea.

I will make a covenant for them on that day,
with the beasts of the field,
With the birds of the air,
and with the things that crawl on the ground.
Bow and sword and war
I will destroy from the land,
and I will let them take their rest in security (2:20).

The customary way of understanding this announcement of salvation has been to see this twofold peace (i.e., with the animal world and with other nations) as a consequence of reconciliation with God. Such an interpretation, which usually places the passage within the context of Mosaic theology, does appreciate the mythic world view of ancient Israel, but it fails to emphasize nature's intrinsic value, a significant feature of Israel' wisdom tradition. The following study will interpret this verse from the latter perspective.

The Framework

Most scholars agree that the Book of Hosea falls easily into two major sections, chapters 1–3 and 4–14, each distinct in length and composition. Although the overriding theme of the book is the covenantal relationship between God and Israel, this relationship is characterized by means of various motifs. The principal motif in the first section (chapters 1–3), the focus of this study, is that of marriage.

Chapters 1 and 3 constitute a kind of framework for chapter 2. While the former describe Hosea's own marriage, the latter uses this description as a metaphor to examine the covenantal relationship between God and the nation. Thus, the marriage motif both links all three chapters to each other and differentiates the first and third chapters (the framework) from the second (the metaphor). These three chapters also differ in genre. The first and third are narratives, while the second is a collection of prophetic speeches. Finally, although chapters 1 and 3 are both narratives, they differ from each other. Chapter 1 is a third-person biographical account of the prophet's marriage. Chapter 3 describes that same relationship in first-person autobiographical speech.

Historical-critical inquiry has tried to explain the narrative discrepancies of these chapters. Was Gomer a harlot before her marriage or did she become one subsequent to it? Was her harlotry a profession or a part of Canaanite cultic performance? Are there one or two wives? Scholars have been unable to arrive at a consensus regarding these questions (Yee 1987, 52–95). However, as important as historical reconstruction may be, it is the biblical text and not the history behind the text that is the primary focus of interpretation, and attention should be given to the structural coherence of the final literary work rather than apparent historical disparity.

In defense of historical investigation, we should note that the narrative sections most likely did originate with Hosea and/or his disciples. His turbulent marriage may well have been a kind of prophecy in action, a way of living out the contrast between the infidelity of one partner (Israel) and the tenacious commitment of the other (God). However, Hosea's words and deeds seem to have had prophetic significance beyond his own original historical moment (Emmerson 1984, 1–8). Subsequent generations, certainly the Exilic community, would regard his message as a metaphor for understanding their own experience of covenantal faithlessness/faithfulness.

The structure of and the symbolic characterizations within the biblical discourse itself encourage such an interpretive shift. The two sign-acts in chapters 1 and 3 constitute a literary framework wherein the historical dissolution and reestablishment of the marriage of the man and woman become the symbol for understanding the rupture and re-creation of the metaphorical marriage of God and the people considered in chapter 2 and, indeed, in the remainder of the book. Since the focus of the Book of Hosea is on the covenantal relationship rather than the conjugal one, the historical details of Hosea's own marriage recede into the background and his prophetic message becomes applicable to other historical situations. What may once have been a prophetic event in the life of an eighth-century prophet now has become a symbolic way of understanding covenantal relationship (Childs 1979, 373–84).

The account of the relationship between God and Israel is found in chapter 2. (The *New American Bible* and the *Jerusalem Bible* conform to the Hebrew numeration, while the *New Revised Version* and the *New English Bible* conform to the Vulgate. The Hebrew numeration will be followed here.) The chapter alternates between words of consolation and those of condemnation, and this makes difficult any attempt to draw logical connections between the chapter's various parts. An interest in historical reconstruction has prompted some contemporary translators to emend the original order of the text. (In the *New American Bible,* 2:1-3 follows 3:5 and 2:8-9 follows 2:15.) This is not only unnecessary but, from a literary point of view, mistaken, for not only the content but the very structure of chapter 2 conveys the message of the prophet, as will be shown. Furthermore, it is the text as received that is canonical and, therefore, the word of God, and not the text reconstructed by interpreters. Logical progression of thought, while undoubtedly a predilection of the Western mind, does not seem to have been the primary concern of ancient Israel. Its world view was much more symbolic than is ours and its forms of discourse were frequently more circular than linear. A brief analysis of the chapter will demonstrate its literary distinction and will show that, with all of its warning and judgment, the Book of Hosea contains a fundamental message of salvation that is communicated as much from its structure as from its content.

The first point of significance is the obvious literary framework that envelops the core of its message:

1-3 intial oracle of salvation
4-17 oracle of judgment
18-22 oracle of salvation
23-25 concluding oracle of salvation

The linguistic and thematic similarities of verses 1-3 and verses 23-25 create an *inclusio* for the story of the relationship between God and the nation (Cassuto 1973, 104-6). Both sections are prophetic announcements of future salvation. They begin with the phrase "and it shall be," a recognized eschatological preface, and they each contain a second prophetic construct. The Hebrew verb forms of verses 1-3 are perfect consecutive, a prophetic way of describing the future as if it were already accomplished, and verse 23 includes the prophetic formula "on that day." Both passages also contain word-play on the name Jezreel and inversions of the names of the other two children, that is, My-People and Compassion.

The eschatological visions described in the two passages, though quite different from each other, depict scenes of restoration and prosperity. ("Eschatological" is used here in the broad sense, referring to a future time when the circumstances of history will be so transformed that one can speak of a new and entirely different state of affairs which unfolds within the framework of history.) In verses 1-3, reference to the increase of the people, as immeasurable as the sands of the sea, recalls the promise made to the ancestors (Gen 22:17; 32:13; cf. 1 Kgs 4:20). The eschatological condition of fulfillment envisioned here is threefold: Judah and Israel will be reunited in one realm, under one head, in their own land. In verses 23-25, Israel's eschatological salvation is described in terms of the transformation of Jezreel from a valley remembered for its infidelity to a land renowned for its fruitfulness. These two passages have a dual function. They serve both as prologue and epilogue for the reversals described in chapter 2 and as oracles of salvation promising a reversal of the conditions portrayed in chapters 1 and 3. It is important to note that the features of this eschatological fulfillment bear the contour of Israel's own unique history, a history founded on specific promises and an exclusive covenant.

Reversals

The centerpiece of chapter 2 consists of two oracles, one of judgment (vv. 4-17) and the other of salvation (vv. 18-22), which together comprise a single literary unit (Brueggemann 1968, 110–23). The first divine speech borrows the form of a lawsuit and consists of an accusation, three indictments, and three judgments against the unfaithful Israel:

4–17 oracle of judgment
- 4 legal appeal
- 6 indictment
- 8 judgment "therefore"
- 10 indictment
- 11 judgment "therefore"
- 15b indictment
- 16 judgment "therefore"

The reversal of the proclamation of marriage ("She is not my wife; I am not her husband," v. 4) is a declaration of divorce and in formulation is similar to the statement made by the name of the second child, Not-My-People (Cassuto 1979, 120–23; Wolff 1979, 33; contra, Andersen and Freedman 80, 220–24; Emmerson 1984, 15; Mays 1969, 36–38; Stuart 1987, 47). Three times evidence of infidelity is reported (vv. 6–7, 10, 15b); three times divine judgment is passed ("therefore," vv. 8, 11, 16). The marriage motif is employed to characterize Israel's sin, which in reality was probably some form of cultic violation. The marriage/family metaphor itself shifts slightly throughout this oracle, sometimes representing the people as the sinful wife, at other times as the mother of sinful children. The amorous language has a dual function, being used first in a sensual sense and then metaphorically.

That God is unwilling to rule out any possibility of reconciliation is clear from the admonition to repentance urged in verse 4b followed by a legal ultimatum in verse 5. However, if Israel will not agree to put an end to its adulterous wanderings, then God will punish the nation either by limiting its ability to wander or by inhibiting the success of its comings and goings (v. 8). The sentence will fit the crime. It is difficult to understand as genuine conversion the return that Israel considers ("I will go back to my first husband"), since it is motivated by deprivation rather than repentance ("for it was better for me then than now," v. 9). The second indictment asserts that Israel did not acknowledge God as the source of fertility, symbolized in the traditional formula "grain, wine, and oil" (v. 10). Instead, it ascribed this bounteous yield, along with wool and flax, to other gods (v. 7). For this reason, God will deprive Israel of both fruits and products, and since there will be no crops, there will be no celebrative harvest festivals. The third and final indictment is similar to the second one. Israel has adorned herself in order to seduce her lovers, thus forgetting God (v. 15b). Since remembering, knowing, and obeying were all aspects of covenant devotion, failure in these activities was seen as infidelity. This third indictment ends with a solemn declaration identifying the pronouncement as a divine oracle: "says YHWH."

The penalties that God imposes are all in some way the antithesis of what Israel envisioned. The first ruling declared that if the nation was unwilling to end its pursuit of other gods, it would be prevented from wandering or would wander to no avail. The second punishment would do more than contain Israel. It would strip the people of the abundant harvests that they already enjoyed but wrongly attributed to the supposed productive power of other gods. Prosperity would be turned into privation. In the third ruling, God announces the consequences that Israel will suffer as a result of its infidelity. Since the divine rulings seem to grow more severe as Israel becomes more recalcitrant, one would expect this third sentence to be the harshest, and the reversal the most dramatic. Here again, the artfulness of the author can be seen, for while the third sentence clearly fits into this pattern of reversal, it does so in such a way as to alter the entire movement of the divine speech. Israel set out to seduce her lovers, and God will respond by seducing Israel (Clines 1979, 83–103).

The content of this judgment is an oracle of salvation which employs but reverses the meaning of some of the prominent imagery from the Exodus tradition: the wilderness; the valley of Achor, through which the people entered Canaan; and the response of each of the covenant partners. First, the wilderness theme functions as an antanaclasis, a figure of speech in which the same word is repeated with a different, even contrary, signification. In the earlier tradition, the wilderness was a desolate place of testing and failure. Here it becomes the trysting place to which Israel is allured, there to embark on a second love affair with God (v. 16). The austerity associated with the first meaning of wilderness here becomes the kind of single-mindedness necessary for the nation to realize that YHWH, and not some other god, is really the source of all life and fertility. A play on words depicts a second reversal. The valley of trouble (the meaning of *Achor,* cf. Josh 7:24-26), the site of deception and punishment, becomes a gate of hope (cf. Andersen and Freedman 1980, 264). Finally, the Exodus came about because YHWH responded to the devotion of the people (cf. Exod 3:7-8). Their deliverance followed their crying out to God. Here the dynamic is the reverse. God saves first and only then do the people respond with devotion. Clearly, the reestablishment of the covenant announced in the Book of Hosea is more than a restoration of what is signified by Exodus theology; it is a creation of something new.

Creation

This divine judgment, which is really a pronouncement of salvation, boasts a chiastic structure:

18 marital theme

20 creation theme

21 marital theme

This framing of the creation theme highlights its centrality. The speech itself begins with the prophetic phrase "And it shall be," adds the eschatological designation "that day," and follows with the oracular designation "says YHWH," all of which signify the importance of the saying. Themes and metaphors that were prominent in the earlier description of Israel's infidelity (husband/wife, the Baals, fertility of the land) are prominent but reinterpreted in verses 18–19, showing that the disintegrated marital relationship will be restored.

Verses 21–22 treat the marital theme in a slightly different fashion. There the verb "espouse" alludes to a premarital status when the groom would pay the bride-wealth to the family of the bride (cf. 2 Sam 3:14). This suggests that in these verses the joining of the man and the woman is something other than a *re*union; it is a new union. The bride-wealth (i.e., the attributes of righteousness, justice, love, and mercy) is the fortune that will be brought to the marriage by the groom to become part of the legacy enjoyed also by the bride. Here too themes found in earlier verses reappear but are changed. For example, two verbs that carry covenantal connotations describe Israel's disposition toward YHWH: first forgetting (vv. 10, 15), then knowing (v. 22). Thus verses 18–19 depict the new relationshp as a restoration, while verses 21–22 present it as a totally new pact.

The heart of the chiasm is verse 20. The phrase "On that day" marks the eschatological character of the saying. Here the language is more formal, the covenant making more obvious, and the frame of reference more universal. This covenant between "them" (most likely the children; see Andersen and Freedman, 1980: 121f.) and "the beasts of the field, the birds of the air, and the things that creep on the ground" is made by God. Although covenants were frequently made between hostile individuals or nations (cf. Gen. 21:27; Josh 9:6-16), they were also made between people who were on friendly, even familiar terms (cf. Gen 31:44; 1 Sam 18:3). Consequently, it should not be presumed that before the making of the covenant, enmity existed between the animals and "them," if such enmity is not stated explicitly in the text. The phrase that names the animals is reminiscent of the creation narrative in Genesis 1:30 as well as its companion passage, the covenant making in Genesis 9:8-10. In neither of these earlier passages is the primary focus on establishing peace between threatening predators and vulnerable human beings. Instead, it is on the harmonious character of creation in its pristine state. It is certainly plausible that the expression carries the same meaning here.

A second phrase, "bow and sword and war," while it partially repeats the listing of actual weapons of human warfare found in 1:7, can also be understood as another link with the Priestly creation tradition. Each one of these words is found in other passages where God is engaged in cosmic battle. The first appears in Genesis 9:13-16. There the *bow* in the sky was an eternal sign of God's promise that never again would the order of creation be overturned. The bow was probably the weapon of the divine warrior, who was victorious over the forces of primeval chaos. This interpretation is supported by several Mesopotamian artifacts depicting arrows in a creator-god's quiver as lightning bolts. Hanging the bow in the sky was a sign that the primeval war was over and all of creation could rest secure. Like the divine rest after creation (Gen 2:2-3), retirement of the bow heralded the establishment of order. The *sword* is also a weapon found in the hand of the creator-god. In an ancient Ugaritic poem recounting the primeval cosmic battle, Leviathan, a serpent with seven heads, is slain by the creator-god. An almost identical passage is found in Isaiah 27:1 where YHWH slays the serpent with a sword (cf. Ps 74:13-14; Job 26:13; in other places the monster is called Rahab, cf. Isa 51:9; Ps 89:10f.). Finally, in one of the earliest pieces of Hebrew poetry, YHWH is called a man of *war* (Exod 15:3).

These examples support the claim that in Hosea 2:20 the weapons of war are not those brandished by the historical enemies of Israel. They are, instead, the weaponry wielded by the divine warrior during the primordial battle. If this interpretation is tenable, one can conclude that here cosmic harmony is not merely a consequence of the covenant between "them" and the animals, but is a constitutive factor of the actual recreation of the covenant. This oracle of eschatological salvation, promising the destruction of the armory, assured Israel of the end of the cosmic battle with chaos, a chaos broader than, but inclusive of, Israel's recalcitrance. Henceforth, there would be no need of weapons. A new covenant had been made, a covenant not dependent upon Israel's fidelity or repentance but upon God's creative love, a covenant with an eschatological character that was cosmic rather than historical.

Relating covenant making or breaking with the establishment or disruption of the created world was not uncommon in the prophetic tradition. Prophets frequently depicted the dire consequences of a breach in the covenant bond as an overturning of the very order of creation (cf. Jer 4:23-26). They also insisted that the relationship established by God after Israel severed the covenant bond would be something quite unprecedented (e.g., new heavens and a new earth, Isa 65:17; 66:22; a new covenant, Jer 31:31; they would be given a new heart and a new spirit, Ezek 11:19; 18:31; 36:26).

Clearly, this oracle of salvation announces a relationship between YHWH and Israel that will be a total reversal of the one depicted in chapters 1 and 3. Furthermore, it is clear that this transformation is not due to Israel's repentance and conversion, but to God's good favor. Finally, and most important for this study, this new covenant bond is more than a restoration of the one severed by Israel's infidelity; it is new creation (Wolff 1979, 55). It is significant that the covenant described here does not focus on the theological issues linked with Israel's specific history of salvation (e.g., Exodus, Torah, etc.). In that tradition, nature was depicted as an instrument of God's blessing and/or punishment (cf. Lev 26; Deut 28). There, its significance and value seemed to hinge on the conditionality of *torah* responsibility. The imagery here recalls the primordial covenant with all of creation, a covenant that exalted nature's intrinsic value and not simply its instrumental value. By returning Israel to the pristine moment of primeval creation, rather than merely the wilderness period of its national origins, this theology underscores the distinctively new relationship that God was establishing. This is a very important point, for it not only grounds God's commitment in the very structures of the natural world but also rules out any kind of anthropocentric imperialism.

A Door of Hope

The revelatory value of the Bible is found primarily in what is announced in the present, not in what was announced in the past. Interpretation is more than just the gathering of information about the text (historical criticism in a narrow sense). It can be defined as the meeting of the world of the reader with the world of the text in such a way that the meaning of the text takes hold of the reader. It is the explanation of the meaning of a text in a new context. This meaning may or may not be the original meaning of the text. Its challenge is in the way it fashions the minds and hearts of believing people, enabling them to live lives of faith and integrity in a world that is ever-changing, not merely in the claims that it makes. In a very real sense, interpretation, like a door, opens the text to new understandings.

A careful reading of the Bible reveals that when Israel was weak and needed reconstituting, the traditions afforded a kind of constitutive support. At other times, when the people seem to have strayed from their fundamental identity as chosen and delivered, the tradition provided a prophetic challenge.

This message from Hosea, which functioned prophetically in one way during the biblical period, functions prophetically in an entirely different way today. Originally, rooting Israel's historical regeneration in creation theology both emphasized the absolute sovereignty of divine initiative and suggested a kind of primordial innocence to the new covenant relationship. The current ecological crisis refocuses our perception of the meaning of this passage. It does not confirm us in our prevailing view of theology, as it did formerly, but challenges us to think anew, in at least three ways. First, its vision of the absolute sovereignty of divine initiative confronts and confounds present-day *hubris* born of our scientific expertise and technological manipulation. Secondly, it calls people of conscience to a kind of primordial innocence in their relationship with the natural world. Finally, it resists any kind of anthropocentric imperialism and, instead, celebrates the intrinsic, not merely the instrumental, value of nature. The conversion that an interpretation of Hosea, such as the one proposed here, calls us to is no less radical than the conversion to which our biblical ancestors were summoned. Whether or not we respond to it in righteousness, justice, love, and mercy (v. 21) remains to be seen.

REFERENCES

Andersen, Francis I., and David N. Freedman
1980 *Hosea.* Anchor Bible 24. Garden City: Doubleday.

Brueggemann, Walter
1968 *Tradition for Crisis: A Study in Hosea.* Richmond: John Knox.

Cassuto, Umberto
1973 *Biblical and Oriental Studies.* Vol. 1: *Bible.* Jerusalem: Magnes Press.

Childs, Brevard S.
1979 *Introduction to the Old Testament as Scripture.* Philadelphia: Fortress.

Clines, David J. A.
1979 "Hosea 2: Structure and Interpretation." *Studia Biblica* 1978, JSOT Supp. 11. Sheffield, England: University of Sheffield.

Emmerson, Grace I.
1984 *Hosea. An Israelite Prophet in Judean Perspective.* JSOT 28. Sheffield, England: University of Sheffield.

Hessel, Dieter T., ed.
1985 *For Creation's Sake.* Philadelphia: The Geneva Press.

McFague, Sallie
1982 *Metaphorical Theology: Models of God in Religious Language.* Philadelphia: Fortress Press, 1982.

Mays, James L.
1969 *Hosea.* The Old Testament Library. Philadelphia: Fortress.

Sanders, James A.
1987 *From Sacred Story to Sacred Text.* Philadelphia: Fortress.

Stuart, Douglas
1987 *Word Biblical Commentary: Hosea-Jonah.* Waco: Word Books.

Tracy, David
1975 *Blessed Rage for Order,* 43–63. Mineapolis: Winston-Seabury.
White, Lynn, Jr.
1967 "The Historical Roots of Our Ecological Crisis." *Science* 155:1203–7.
Wolff, Hans Walter
1979 *Hosea.* Philadelphia: Fortress.
Yee, Gale A.
1987 *Composition and Tradition in the Book of Hosea: A Redaction Critical Investigation.* Atlanta: Scholars Press.

2

Paul for the Ecozoic Age

Barbara Reid, O.P.

We are at the end of the Cenozoic age, the era of geological history that began some 65 million years ago, and are moving into the Ecozoic age. So claims cultural historian Thomas Berry, C.P., who is one of the prominent voices articulating the relation between ecology and theology.[1] The new Ecozoic age envisioned by Berry, who coined the term, would be an era of mutually enhancing human-earth relationships. For this to occur, a revolution in our understandings of God, the earth, time, human beings, other species, and all human institutions must be effected.

This present moment of crisis and transition also demands new interpretations of the biblical tradition. In what follows, we will describe some of the conditions and characteristics that Berry sees as necessary of the Ecozoic age. In this context, we will highlight some of the writings of the Apostle Paul and suggest ways in which these might guide us into the approaching era.

The Present Situation

Berry observes that what is happening in our times is the most momentous change in the four and a half billion years of the history of the Earth. It is not just another historical transition or cultural change. The current devastation of the planet by human beings is altering all former human modalities of being. Can Christian thought deal with what faces us today? Berry says that it cannot do so out of its existing resources. But he observes that neither can we deal with these problems without our religious traditions. Berry asserts that we cannot deductively get our guidance for today from the past. But there is a new revelatory experience of the divine through our present understanding of the universe, which can give us a new sense of being Christian (pp. 6–7). Even so, he advises, Christian thought must change in a way in which it has never changed before.

The Divine in a Time-Developmental Universe

When Berry begins to talk about God, he recognizes the inextricable link between God and creation. One might try to talk about God as being prior to or independent of creation, but, he says, "in actual fact there is no such being as God without creation." Berry defines "God" as "the ultimate mystery of things, something beyond that which we can understand adequately." The divine is "the ineffable, pervasive presence in the world about us" (pp. 10–11).

In our new historical context this divine presence in creation is understood differently than in times past. Berry describes how, originally, the divine was perceived as manifested throughout the natural world. Time was thought to move in ever-renewing, seasonal cycles of change. The universe was believed to have always existed as it was; so it always would be. No actions by human beings could alter it.

In biblical tradition, there emerges a new sense of history, in which the universe is thought to have come into being at a particular moment. Time is still thought of as a cycle of seasons, although in terms of humans, it is developmental.

Today, we share the biblical consciousness that the universe had a beginning in time. But unlike the biblical writers, who thought the cosmos was created once and for all, we know today that the universe is emergent, undergoing irreversible sequences of transformations over billions of years. The challenge for Christians, says Berry, is to move into a sense of developmental time as sacred time. This requires a monumental change: from perceiving the world as *cosmos,* to perceiving it as *cosmogenesis.* That is, our notion of the world must move from thinking of a world that *is* to an understanding of a world that is *becoming.* It is with this consciousness that Christian teachings must be reinterpreted, if Christianity is to survive in any effective manner (p. 74).

Conditions for the Ecozoic Age

Berry lists six conditions of survival in the world ahead of us. In each, he avows, the religious dimension must enter in and help to reshape our consciousness. First, "the universe is a communion of subjects, not a collection of objects" (p. 96). Without this perspective, which respects the interior dignity of all things on the globe, we witness the plundering of the planet by industrial-commercial society.

Second, "the earth exists and can survive only in its integral functioning" (p. 96). The earth is a single reality and cannot be preserved in fragments.

Third, "the earth is a one-time endowment" (p. 96). There is no way to reverse the damage if we destroy the earth. There is no second chance.

Fourth, we must realign all human institutions to reflect a consciousness that the earth is primary and that humanity is derivative. Human beings can only be saved within the earth community. It is absurd, for example, says Berry, to choose an expanding human economy that diminishes the earth economy. Without the latter, there is no context for the former (p. 97).

Fifth, the entire pattern of the earth's functioning is altered in the transition from the Cenozoic to the Ecozoic age. Humans had nothing to do with the emergence and formation of the Cenozoic era. But in the Ecozoic period, nature will not function without its acceptance, protection, and fostering by humankind. This is a vast change in the way the biosystems of the planet function and places new responsibilities on human beings. People will need to be committed not only to taking care of things as they are at present, but also to enabling them to be what they are called to be. The earth never remains the same, but is in a process of continuing transformation (pp. 98–99).

Sixth, we need new ethical principles which recognize the absolute evils of biocide, the killing of life systems themselves, and geocide, the killing of the planet (p. 100).

Paul and the Meeting of the Ages

We are not without precedent in Christianity for radical changes of consciousness and reinterpretation of biblical tradition. Although Berry claims that human beings have never faced change of the magnitude that they do today, in every age the Christian task has been to discern the continuity of the tradition and to reinterpret it in terms of the changed situation of the day. One New Testament author who is helpful on this point is Paul, who speaks of the change demanded by the Christ event. He also considers his day to be a confluence of the "ends of the ages" (1 Cor 10:11).

From his Pharisaic background, Paul viewed salvation history as divided into three eras: (1) from creation to Moses, when there was no Law (Rom 5:13-14); (2) the period of the Law, from Moses to the Messiah; and (3) the messianic age. Having come to believe in Christ as the Messiah, Paul asserts that "Christ is the end of the Law" (Rom 10:4) and that with Christ the messianic age is inaugurated. This new age demanded a radical change in the way salvation was understood.

New Creation

Twice Paul speaks of this new age of salvation as "new creation." The first instance occurs in the conclusion of his fiery letter to the Churches of Galatia (Gal 6:15). He is most distressed with this community, which has been swayed by other evangelizers, preaching a different message from the one they received from him. These others advocated circumcision and the observance of Jewish dietary regulations for the newly converted Galatian Christians. Paul is aghast that they have so easily been persuaded by this "different gospel" (Gal 1:6). He argues forcefully that "a person is not justified by works of the law but through faith in Jesus Christ" (Gal 2:16). In his final appeal, Paul asserts, "neither circumcision nor uncircumcision is anything; but a new creation is everything!" (Gal 6:15).[2]

In using the language of "new creation," Paul indicates the magnitude of change signalled by the new era begun with the Christ event. He writes that this "new creation" means that there is a new and exclusive way of salvation: the way of the cross of Christ. This cannot be combined with the former way of the Law and circumcision. That this change in perspective was most difficult for Paul's converts is evident from the number of times he discusses the issue. Justification by faith and not by works of the Law is a major theme in his letters to the Churches of Galatia, Philippi, and Rome.

The second time Paul uses the expression "new creation" is in the context of his reflections on the ministry of reconciliation in 2 Corinthians 5:17. This verse can be rendered either "Whoever is in Christ is a new creation" (NAB) or "If anyone is in Christ, there is a new creation" (NRSV). Paul's statement allows for the interpretation that not only the individual Christian but the objective situation are transformed: "everything old has passed away" and "everything has become new" (2 Cor 5:17). In a sense, Paul's assertion reflects the same comprehensiveness of change that Berry articulates.

The Body of Christ

Paul's concept of community could be of help for the formation of the Christian imagination in the Ecozoic age, particularly as he uses the image of the body in Romans 12:3-8 and 1 Corinthians 12:12-26. Paul says, "For just as the body is one and has many members, and all the members of the body, though many, are one body, so it is with Christ" (1 Cor 12:12). He goes on to describe how each individual member has a part to play for the proper functioning of the whole: the foot, the hand, the ear, the eye. He describes how God has so arranged the body that "there may be no dissension within the body, but the members may have the same care for one another. If one member suffers, all suffer together with it; if one member is honored, all rejoice together with it" (1 Cor 12:25-26).

The origin of this figure in Paul is disputed, but most hold that is comes from contemporary Hellenistic notions about the state as the body politic. This idea is found as early as Aristotle and later became part of Stoic philosophy. Paul's home town, Tarsus, was a renowned center of Stoicism. Moreover, the image occurs in a letter to a community that is torn apart by divisions (1 Cor 1:10-17).

Paul has in mind the cooperation of humans, but we might expand this image to envision the first two of Berry's conditions for the Ecozoic age: perceiving the universe as a communion of subjects, not a collection of objects, that function integrally in our planet. We might interpret Paul's idea, "if one member suffers, all suffer together with it" (1 Cor 12:26), not only in terms of human members of the planet, but of all forms of life on the earth. For example, we could paraphrase, "if the ozone layer suffers, plants suffer, and human beings suffer with it."

The Cosmic Christ

Another Pauline image that may deepen the Christian commitment to the Ecozoic age is that of the cosmic significance of Christ, as found in the letters to the Colossians and Ephesians. The early Christian hymn (Col 1:15-20) incorporated into the first chapter of the Letter to the Colossians proclaims Christ as an agent in creation and describes the cosmic impact of the reconciliation effected by him: "in him all things in heaven and on earth were created, things visible and invisible . . . all things have been created through him and for him. He himself is before all things, and in him all things hold together . . . through him God was pleased to reconcile to himself all things, whether on earth or in heaven, by making peace through the blood of his cross." A similar description of the cosmic sovereignty of Christ is found in Ephesians 1:22-23.

As Berry notes, for the Ecozoic age, Christians need to move from an overemphasis on the individual Jesus to a sense of the Christ dimension that has been present in the universe from its inception (p. 77). What enables us to speak of this Christ dimension is that everything has both its individual phase and its cosmic phase. As Berry explains, "everything that has happened in the whole course of the universe is present in each one of us, just like every atom is in contact with and affects every other atom of the universe" (p. 78).

Ethical Demands

Berry speaks in his last two conditions for survival of the responsibilities of human beings in the Ecozoic age and of the need for new ethical principles. Here, we find that Paul's outlook is less helpful. Steeped in an apocalyptic mindset, we find him saying, "I think that, in view of the impending crisis, it is well for you to remain as you are" (1 Cor 7:26), when addressing questions about marital and social status. Paul expected the parousia imminently, as we can see from his declarations, "time is running out" (1 Cor 7:29), and "the world in its present form is passing away" (1 Cor 7:31). Because of this he counselled the Corinthians to do nothing to change their social status (slavery or freedom), their circumcision, or their marital situation.

Paul could be understood to advocate an escapist stance toward the world; that is, since the end is near, any effort to change the present earthly situation is useless. One waits, instead, for God's intervention to right all wrongs for the just. In Berry's vision of the Ecozoic order, humanity cannot take a passive role. The steps toward the destruction of the planet we have already taken demand the active nurturance of nature by human beings.

Continuing Transformation

Berry argues that we must not only take care of what is but enable everything to develop into what it is called to become in the process of continuing transformation. We find the idea of continuing transformation for the believer in 2 Corinthians 3:18, where Paul says, "all of us, with unveiled faces, seeing the glory of the Lord as though reflected in a mirror, are being transformed into the same image from one degree of glory to another." A notion of ongoing transformation of the universe itself would have been alien to Paul, but that he has some idea of a transformation of creation is evident in Romans 8:18-30, where he sketches his vision of the glorious destiny of believers. He speaks not only of the transformation of the person of faith, but also that "creation itself will be set free from its bondage to decay and will obtain the freedom of the glory of the children of God" (8:21). Paul ascribes to creation the same ends and desires as human creatures. As humanity waits (8:23, 25) in hope (8:24-25), groaning for the fullness of redemption (8:23), so creation awaits (8:19) in hope (8:20) groaning (8:22) to be set free (8:21).

Another aspect that Berry would have us stress is the inherent relationship between the human and the divine within the natural order. In one sense, Paul does this in Romans 1:20, where he asserts, "Ever since the creation of the world [God's] eternal power and divine nature, invisible though they are, have been understood and seen through the things he has made." This, Paul says, should move even those who do not believe in Israel's God to praise the Creator (Rom 1:21). But they do not. Nor do the Jews, who have the Law (Rom 2:1-29). For Paul, neither the pervasive presence of the divine in creation nor the covenant relationship with Israel was able to bring humankind into right relation with the Creator. As he sees it, this was accomplished only by a new act of creation by God: the Christ event (Rom 7:25).

Sacrifice

One important thing to which Berry alerts us is that the reshaping of all aspects of our existence for the Ecozoic age cannot occur without our willingness to accept sacrifices. He explains, "Sacrifice is the idea that whatever is achieved has a price" (p. 133). There is always give and take. Berry identifies the root of the tragedy of our times as our unwillingness to make return for what has been given us. We take from the earth for our comforts but are not willing to give back to it. In this, human beings deny the reality that all life is reciprocal. "Every living being is sacrificed for other living beings . . . everything feeds on other beings and nourishes other beings" (p. 134).

Another way Berry explains the sacrificial mode is in terms of the different selves that we have. "We have our personal self, our family self, our earth self, and our universe self. . . . Sacrifice, ultimately, is the choice of the larger self, because when the larger self is endangered by the smaller self, the smaller self must give way to the larger self when it is in its authentic mode" (p. 135).

Here is where New Testament theology could make the most significant contribution for the Ecozoic age. The sacrifice of the life of Christ on the cross for the life of all is the essence of the Christian message. Central to Paul is the theology of the cross. Although Paul's language is different from Berry's, the thought of the smaller self being sacrificed to the larger self could be seen in a number of Pauline texts. In Philippians 2:3-11, Paul introduces an early Christian hymn with an admonition,

> Do nothing from selfish ambition or conceit,
> but in humility regard others as better than yourselves.
> Let each of you look not to your own interests,
> but to the interests of others.
> Let the same mind be in you
> that was in Christ Jesus,
> who, though he was in the form of God,
> did not regard equality with God
> as something to be exploited,
> but emptied himself,
> taking the form of a slave,
> being born in human likeness.
> And being found in human form,
> he humbled himself
> and became obedient to the point of death —
> even death on a cross.

A similar passage is 2 Corinthians 8:9, "For you know the generous act of our Lord Jesus Christ, that though he was rich, yet for your sakes he became poor, so that by his poverty you might become rich."

For some, this will sound like nonsense. But for survival into the Ecozoic age, we must resonate with Paul, who says in 1 Corinthians 1:18, "the message about the cross is foolishness to those who are perishing, but to us who are being saved it is the power of God." If we do not strengthen our inner capacity to act in the sacrificial mode, the alternative, says Berry, is a destructive, technozoic mode in which we attempt to overcome the limits and rhythms and conditions of the planet with technology (p. 116).

Energetic Joy

Finally, lest the enormity of the changes and sacrifices facing us for the Ecozoic age paralyze us with fear, Berry notes that we must enter the birth of new modes of being with attraction, joy, and delight, because a supreme opportunity is being offered us (p. 135). He identifies the gift of delight in existence as one of the things that religion can give us. Such delight is the source of immense energy (p. 110). We might then pray with Paul,

> Rejoice in the Lord always; again I will say, Rejoice. Let your gentleness be known to everyone. The Lord is near. Do not worry about anything, but in everything by prayer and supplication with thanksgiving let your requests be made known to God. And the peace of God, which surpasses all understanding, will guard your hearts and your minds in Christ Jesus (Phil 4:4-7).

NOTES

[1] The observations of Berry that will be examined in this essay are from Berry and Clarke, 1991. Page references throughout refer to this work.

[2] All Scripture quotations are from the *New Revised Standard Version* (New York and Oxford: Oxford University Press, 1989).

REFERENCES

Berry, Thomas, and Thomas Clarke

1991 *Befriending the Earth. A Theology of Reconciliation between Humans and the Earth.* Mystic, Conn.: Twenty-Third Publications.

Fitzmyer, Joseph A.

1990 "Pauline Theology." *New Jerome Biblical Commentary,* 82. Englewood Cliffs, N.J.: Prentice-Hall.

McFague, Sallie
1987 *Models of God. Theology for an Ecological Nuclear Age.* Philadelphia: Fortress.

Reilly, William K.
1991 "Theology and Ecology: A Confluence of Interests." *New Theology Review* 4:15-25.

Part II

Ethical Reflections

3

The Roman Catholic Social Tradition and the Question of Ecology

Thomas A. Nairn, O.F.M.

When Pope John Paul II dedicated his 1990 World Day of Peace message to the ecological crisis, he received high praise. Some commentators at the time hailed the message as being a long overdue Catholic entrance into the issue of the environment. The truth, however, is that the Pope's message was not the Church's first word on the environment, nor was the scope of the message radically new. On the contrary, it was rather another development in the Church's hundred-year-old social tradition, beginning with Pope Leo XIII's great social encyclical, *Rerum novarum* (On the Condition of the Working Class), issued in May of 1891, and continuing through Pope John Paul's encyclical, *Centesimus annus* (One Hundred Years of Catholic Social Teaching), issued in May of 1991.

As regards ecology, this traditon may be seen as having at least three different, though related, phases: the first from Leo XIII until the beginning of the Second Vatican Council, the second from the council through the pontificate of Pope Paul VI, and the third during the present pontificate of John Paul II. Although these phases are in fact different, they revolve around two common rubrics: in each, the Church situates the question of the environment within the larger context of justice and the common good.

First Period: Natural Order

Thirteen years before the promulgation of Pope Leo XIII's great social encyclical, *Rerum novarum,* the Pope published another encyclical, *Aeterni Patris,* in which he proclaimed the philosophy of St. Thomas Aquinas, especially as appropriated by neo-Thomism, as the "received tradition" of the Catholic Church. This philosophy is the foundation upon which the early period of the Church's contemporary social tradition hinged. Dependent upon a particular understanding of natural law, taken to be true in both personal and social ethics, this school maintained that principles of natural law were created by God as perfectly ordered, absolute, and necessary.

The notion of order then became the basis for the articulation of a particular perception of the common good. Defining order as "unity arising from the harmonious arrangement of many objects," papal thinking of this time, as exemplified by Pope Pius XI's *Quadragesimo anno,* emphasized that "a true, genuine social order demands that the various members of a society be united together by some strong bond. This unifying force is present . . . in that common good, to achieve which all industries and professions together ought, each to the best of its ability, to cooperate amicably" (par. 84). Working together toward the common good would result in that harmony and order which God had ordained in society.

Both natural order and the current social order were seen as part of God's plan. On the social level, although the social encyclicals of this period were concerned about alleviating the misery of the poor, they also accepted the social hierarchy of the time. *Rerum novarum* reminded its readers that "there are truly very great and very many natural differences among men. . . . and unequal fortune follows of itself upon necessary inequality in respect to these endowments. And clearly this condition of things is adapted to benefit both individuals and the community" (par. 26). *Quadragesimo anno,* trying to combat the communism of its day, likewise never suggested radical revision of the social division of society. Rather it maintained that the most pressing duty of society was "to get rid of conflict between 'classes' with divergent interests and to foster and promote harmony between the various 'ranks' and groupings of society" (par. 81).

The writings of this period also understood the natural order as ordained by God in a similar hierarchical fashion, with humanity at the pinnacle. Aristotle had already articulated this hierarchy three hundred years before the birth of Christ: "Plants exist to give food to animals, and animals to give food to humans: domestic animals for their use and food, wild ones, in most cases, if not in all, furnish food and other conveniences, such as clothing and various tools. Since nature makes nothing purposeless or in vain, all animals must have been made by nature for the sake of humans" (Politics, 1256b). St. Thomas Aquinas, concurring with this notion, added that "this is proved by the order of Divine Providence which always governs inferior things by the superior. Thus, as humanity, created in the image of God, is above other animals, these are rightly in its government" (Summa Theologica 1.96.1). The Church understood this natural hierarchy as even more basic than its social counterpart. Thus the tradition concluded that only certain creatures, namely humans, had intrinsic value, while all others had mere instrumental value, to be used by and for the good of humans.

Given these analytic tools, it should be clear that this first period could not have many resources with which to develop a constructive ecological ethic. Furthermore, grounding its judgments in the rubrics of order and the common good (itself understood in terms of harmony), this phase displays deep ambivalence regarding ecological questions. On the one hand, humans, being the only creatures with intrinsic value, have the right to fashion nature according to their own ends. *Rerum novarum,* for example, speaks that whatever one owns, "whether the property be land or movable goods," should be "as completely at man's disposal as the wages he receives for his labor" (par. 9). Forty years later, Pius XI would attempt to limit this right of ownership. Nevertheless, he still adds that the "putting of one's own possessions to proper use, however, does not fall under [commutative] justice, but under certain other virtues and therefore it is 'a duty not enforced by the courts of justice' " (par. 47). The very end of nature is to be used by humanity.

On the other hand, these same notions of order and the common good were used in the tradition as limiting concepts. The tradition attempted to provide natural law boundaries beyond which humans ought not transgress. Thus *Quadragesimo anno* states that "nature, rather the Creator himself, has given man the right of private ownership not only that individuals may be able to provide for themselves and their families but also that the goods which the Creator destined for the entire family of mankind may through this institution truly serve this purpose. All of this can be achieved in no wise except through the maintenance of a certain and definite order" (par. 46). This first period thus speaks of the relation of the natural creation to humanity. Unfortunately, this period's understanding of order and the common good, upon which this relation was based, proved insufficient to ground adequately a fully developed ecological ethic.

Second Period: Human Interdependence

For one accustomed to the Catholic vision of nature and society in terms of that order described above, the Second Vatican Council would come as a bombshell. In place of the concept of natural law, understood in terms of order, harmony, and perfection, the council's Pastoral Constitution on the Church in the Modern World, *Gaudium et spes,* developed a different point of departure. When it discussed "particularly urgent needs characterizing the present age," it did so "in light of the gospel and of human experience" (par. 46). This formulation pointed toward a new manner of ethical reflection, replacing a rigid form of natural law. It is within this framework of gospel and human experience that the council would ask Catholics to acknowledge that the Holy Spirit is active not only within the Church but also within the world itself and direct them to thus reflect upon the "signs of the times." The council thus moved from an abstract notion of order and took as its starting point the actual concrete historical experience of men and women.

The idea of the "signs of the times" was not original to Vatican II. First used in Pope John XXIII's 1961 Christmas message, it became an organizing principle of his encyclical *Pacem in terris*. Most basically, the driving force behind such language was the belief that the Holy Spirit is speaking in the day-to-day events of people's lives. The human person is always a being in the world, and therefore history itself must be part of what it means to be human. Once history enters into Christian reflection, the faithful are invited to analyze the changes around them and see how they affect the human condition itself. The events of the world therefore have a major importance for the Church. The "times" themselves are able to teach the Church (see pars. 16, 44). In one sense, *Gaudium et spes* leads one to a remarkable sense of divine providence and God's actual care for the world and its people.

By means of the language of "signs of the times" the constitution challenges the Church to see the deep connection between the actual events of the world, both the good and the bad, and the Word of God. Taking its prompting from the eighth chapter of St. Paul's Letter to the Romans, the council summons all believers to see that the world is groaning and in the pangs of childbirth, awaiting its fulfillment in Christ.

It would seem that the imagery of "signs of the times" and a world in the pangs of childbirth would be a more than apt vision upon which to base an ecological theology. Yet, although *Gaudium et spes* does speak about environmental issues, the reasoning of the council is not so straightforward. In a move paralleling that of the first period, the council uses the notion of "signs of the times" and the biblical foundation in which it is grounded to move toward an expanded concept of the common good. The council maintains that it is only in relationship to all the daughters and sons of God that one is an image of God who is Trinity. Relationship to others in love and justice is essential to the divine image in humanity (see par. 24). The council thus moves beyond an understanding of the common good as a harmonious ordering of society to a more interrelational conception, defining it as "the sum of those conditions of social life which allow social groups and their individual members relatively thorough and ready access to their own fulfillment" (par. 26).

Although the context changes, it seems that the evaluation of material creation in relation to humans does not. Discussing the biblical theme of "a new heaven and a new earth," the document concludes: "with death overcome, the sons of God will be raised up in Christ. . . . While charity and its fruits endure, all that creation *which God made on man's account* will be unchained from the bondage of vanity" (par. 39, my emphasis). The natural hierarchy, analyzed in the first period in terms of natural order, remains, although it now reflects this revised understanding of the common good. The council maintains the instrumental value of the non-human world.

Gaudium et spes does speak of ecology. Yet its very placement within the chapter devoted to the economic life is indicative:

> God intended the earth and all that it contains for the use of every human being and people. Thus, as all follow justice and unite in charity, created goods should abound for them on a reasonable basis. Whatever the forms of ownership may be, as adapted to the legitimate institutions of people according to diverse and changeable circumstances, attention must always be paid to the universal purpose for which created goods are meant. In using them, therefore, a person should regard his lawful possessions not merely as his own but also as common property in the sense that they should accrue to the benefit of not only himself but of others. (Par. 69)

Thus the constitution still analyzes the environment and ecological questions in terms of *human* interdependence and its revised understanding of the common good. In doing so, it retains the ambivalence discussed in the previous section.

In the decade which followed the council, the Vatican issued statements on ecology with increasing frequency. In 1967, Paul VI used his encyclical *Populorum progressio* to expand the notion of common good into a planetary phenomenon. He continued this theme in his 1971 apostolic letter, *Octogesima adveniens,* but with a change in emphasis. The earlier encyclical had as a major theme that of "development." It saw the problems of the Third World as solvable by applying the process which had already taken place in the technologically advanced nations to the so-called developing countries. By the time of the apostolic letter, however, the Pope was much more ambivalent regarding the real possibilities of such progress (see par. 41). He began to acknowledge that conditions within the First World nations themselves seemed to prevent the full human development of poorer nations. He stressed that the planetary common good demands global justice; global interdependence entails global responsibility.

Within this context, the apostolic letter cited a number of social problems, from urbanization, through the variety of questions centering around employment, to emigration. In addressing these needs, he saw the role of the Church as "to take part in action and to spread, with real care for service and effectiveness, the energies of the Gospel'' (par. 48). Christians have responsibilities with respect to every social problem. One such area of responsibility is ecology. Pope Paul's ambivalence regarding ''progress'' becomes evident in the paragraph he addresses to this theme:

> Man is suddenly becoming aware that by an ill-considered exploitation of nature, he risks destroying it and becoming in his turn the victim of this degradation. Not only is the material environment becoming a permanent menace—pollution and refuse, new illnesses and absolute destructive capacity—but the human framework is no longer under man's control, thus creating an environment for tomorrow which may well be intolerable. This is a wide-ranging social problem which concerns the entire human family.
>
> The Christian must turn to these new perceptions in order to take on responsibility, together with the rest of men, for a destiny which from now on is shared by all. (Par. 21)

Similarly, the 1971 synodal document ''Justice in the World'' places the question of the environment in the context of global justice:

> It is impossible to see what right the richer nations have to keep up their claim to increase their own material demands, if the consequence is either that others remain in misery or that the danger of destroying the very physical foundations of life on earth is precipitated. Those who are already rich are bound to accept a less material way of life, with less waste, in order to avoid the destruction of the heritage which they are obliged by absolute justice to share with all other members of the human race. (Par. 70, see also pars. 11 and 12)

This discussion of ecology becomes in reality a discussion of responsibilities among peoples, of richer nations toward poorer ones; it is actually a discussion of the revised understanding of a planetary common good.

Two other documents continue this theme of viewing ecology from the perspective of global justice: Paul VI's 1972 message to the United Nations Conference on the Environment and his 1977 message for the World Day on the Environment. The first document begins by stressing the mutual dependence between humanity and the environment and alludes to the contemplative dimension of humanity's relation to the environment, thus giving rise to an expectation for a new basis from which to view ecology. However, the document soon returns to the discussion of a global social justice, similar to that articulated above: ''The compression of distance by progress in communication, the formation of always closer bonds between peoples through economic development, the increasing domination of nature's forces by science and technology, the multiplication of human relationships across the frontiers of nations and race—all of these are, for better or for worse, in hope or toward disaster, elements of interdependence. . . . From now on, with interdependence must go mutual responsibility, for a community headed into the future must be united'' (p. 76). The document then places the question of ecology almost entirely within a context of human interdependence: ''. . . the worst pollution is human misery. . . . Therefore, instead of understanding the fight for a better environment as a fearful reaction of the rich, we should see it as for the advantage of all, an affirmation of faith and hope in its destiny on the part of the human family united in a common enterprise'' (p. 77).

Similarly, Paul VI's 1977 message for the World Day of the Environment offers no new analysis. As with previous documents, it calls for ''a change of mentality, for a conversion of attitude and of practice so that the rich willingly use less and share the earth's goods more widely and more wisely.'' It calls for ''a universal sense of solidarity in which each person and every nation plays its proper and interdependent role to ensure an ecologically sound environment for people today as well as for future generations'' (p. 11). The Church maintains its ecological ethic as subordinate to an ethic of global justice among all peoples.

This second period does move beyond the limits of the first. It does begin to address the environment as an issue important in itself and, in doing so, takes the environment more seriously. Yet the context within which it faces this issue is itself problematic. The Church sees the problem as one of economic and social inequalities among persons, and its cure is to effect a change in *human* social relationships based on solidarity and the dignity of the poor. Humanity remains the apex of creation. The Church unwittingly gives rise to a further conflict, that between the development of peoples and ecological awareness. When viewing the situation from this framework, some will describe the ecological crisis as a ''pseudo-crisis,'' one which keeps us from dealing with the real issues of justice toward the Third World and true development among the poor in our own nation. The environment becomes a ''middle class question'' which keeps us from dealing with the real issues of social justice. This second phase of the tradition's attempt to address the issue of the environment remains almost as unsatisfying as the first. Developing an ecological ethic on the basis of the planetary common good and the notion of global justice does not yet provide the resources for a more complete ethical analysis.

Third Period: Human Co-Creation

John Paul II has referred to the question of ecology in several of his writings. In doing so, he has maintained the tradition's attention to justice and the global common good but has also created a dialectic between this traditional emphasis and a second emphasis, that of contemplation. He brings the two together by means of his rubric "co-creation."

Much has been written about the distinction between the Pope's social ethics and his sexual ethics. Many have contrasted the two, and, depending upon their own liberal or conservative bias, have generally reviewed one area of analysis favorably while being critical of the other. They then try to account for what they regard as inconsistencies within the Pope's thought. This rather common assessment is simplistic at best.

The notion of human co-creation is the common thread which unites these two strands of papal ethical thought. Whether the Pope is speaking of human work, as in his encyclical *Laborem exercens,* or of human love and sexuality, as in his apostolic exhortation *Familiaris consortio* (and the catechesis on human sexuality he undertook by means of his general audiences devoted to this subject), he begins by turning to the first chapters of the Book of Genesis to guide his reflection (see 1981a, pars. 1 and 4; 1981b, par. 10ff.). In both analyses, the notion of the human person as the image of God is dominant. In the encyclical, for example, the Pope acknowledges that "man is the image of God partly through the mandate received from his creator to subdue, to dominate, the earth. In carrying out this mandate, man, every human being, reflects the very action of the creator of the universe" (par. 4). Similarly, his apostolic exhortation contends that "creating the human race in his own image and continually keeping it in being, God inscribed in the humanity of man and woman the vocation, and thus the capacity and responsibility of love and communion" (par. 11). In both documents, the Pope then moves from this understanding of humanity as image of God to an emphasis on human co-creation with God. This concept of co-creation in turn revolves around two axes: subduing the earth and mastering the self.

These two foci are not entirely coordinated in the Pope's thought and reflect a certain ambivalence. There are times when his discussion of humanity's dominion of the earth is quite enthusiastic. Paragraphs four and five of *Laborem exercens,* for example, contain a tremendously positive evaluation of human work in terms of "subduing the earth." Within this context, he praises contemporary industrial society along with its technology. Indeed, the Pope calls technology "the ally of work" (par. 5). Similarly, his apostolic exhortation on the family acknowledges that scientific and technological progress "offers hope of creating a new and better humanity" (par. 30).

Yet parallel to this is the theme of self-mastery. The Pope is not as laudatory here. In his general audience of August 22, 1987, he explained the need to maintain an adequate relationship between the domination of the forces of nature and self-mastery. Self-mastery becomes the subject of his sexual ethics in his praise for that form of self-control necessary for the practice of natural family planning. Self-mastery also enters into his social ethics by means of his description of the subjective nature of work:

> Work is a good thing for man—a good thing for his humanity—because through work man not only transforms nature . . . but also achieves fulfillment as a human being and indeed in a sense becomes "more a human being." . . . All this pleads in favor of the moral obligation to link industriousness as a virtue with the social order of work, which will enable man to become in work "more a human being" and not be degraded by it . . . especially through damage to the dignity and subjectivity that are proper to him. (1981a, par. 9)

There is indeed an Augustinian strain to the Pope's thought which overshadows this discussion: Selfishness and pride also rule the human spirit. Through their actions, humans create themselves either for good or for ill. Technology is not only a source of hope; it can also be used to fuel that human self-seeking which gives rise to evils such as materialism and consumerism. The Pope concludes that "the ultimate reason for these mentalities is the absence in people's hearts of God" (1981b, par. 30).

It is interesting to see that in discussing the movement from selfishness to self-mastery the Pope connects the dominant themes of the previous periods. He speaks of that order designed by God to which humans must submit in their mastering of self, reminiscent of the first period discussed above; and he also maintains the importance of human solidarity, reminiscent of the second period.

The Pope uses these themes as resources as he develops his teaching regarding ecology. His very first encyclical, *Redemptor hominis,* makes use of the previous period's "signs of the times" imagery in his discussion relating humanity and the environment, but it retains an Augustinian ambivalence regarding human possibilities:

> The man of today seems ever to be under threat from what he produces, that is to say from the result of the work of his hands and, even more so, of the work of his intellect and the tendencies of his will. All too soon, and often in an unforeseeable way, what this manifold activity of man yields is not only subjected to *alienation,* in the sense that it is simply taken away from the person who produces it, but rather it turns against man himself, at least in part, through the indirect consequences of its effects returning on himself. (Par. 15)

He concludes:

> Yet it was the Creator's will that man should communicate with nature as an intelligent and noble *master* and *guardian,* and not as a heedless *exploiter* and *destroyer.* . . . The question keeps coming back with regard to what

> is most essential—whether in the context of this progress man, as man, is becoming truly better, that is to say more mature spiritually, more aware of the dignity of his humanity, more responsible, more open to others, especially the neediest and the weakest, and readier to give and to aid all. (Par. 15)

The Pope's starting point is different from that of others before him. Yet his conclusion is amazingly similar. The distinction between humanity and the rest of creation remains. Creation is made for humanity. As described in the encyclical, a major problem of the ecological crisis is that in exploiting creation humans become less human *to one another,* less able to respond to the weak and vulnerable in their midst.

This sort of analysis continues in his 1990 World Day of Peace address. "When man turns his back on the Creator's plan, he provokes a disorder which has inevitable repercussions on the rest of the created order. If man is not at peace with God, then earth itself cannot be at peace" (par. 5). The Pope's solution to this crisis lies in the acceptance of the virtue of self-mastery and his correlation between the notion of natural order and that of interdependence:

> The earth is ultimately a common heritage, the fruits of which are for the benefit of all. . . . This has direct consequences for the problem at hand. It is manifestly unjust that a privileged few should continue to accumulate excess goods, squandering available resources, while masses of people are living in conditions of misery at the very lowest level of subsistence. Today the dramatic threat of ecological breakdown is teaching us the extent to which greed and selfishness—both individual and collective—are contrary to the order of creation, an order which is characterized by mutual interdependence. (Par. 8)

His message, beginning with an analysis of human sin and selfishness, moves to a consideration of that order designed by God and an equation of that order with the postconciliar understanding of human interdependence. As in the previous period, the environment problem is understood in terms of human interdependence.

Yet another element is present in Pope John Paul's analysis of co-creation in general and of his discussion of ecology in particular, an element which is more suggestive than fully articulated. Developing a topic introduced by Paul VI in his 1972 address, Pope John Paul's 1990 World Day of Peace message uses the language of contemplation in assessing an environmental ethic. This call to contemplation becomes an invitation to deepen one's attitude of respect toward life into a respect for the integrity of the entire ecosystem. Most particularly, it calls one to move beyond the facile position that only humans have intrinsic value while the rest of material creation has instrumental value, created for human use.

At present, however, this element of contemplation is not developed enough to offer an adequate corrective to the anthropocentrism of the Church's recent tradition of social ethics.

Conclusion: The Limits and Possibilities of the Tradition

When one looks at the Church's contemporary social tradition as described above, one sees certain constants within all three phases: the centrality of the notion of common good, a corresponding notion of harmony and order, human interdependence, a distinction between the instrumental value of material creation and the intrinsic value of humanity, and a setting of the ecological question within that of human society. In spite of differences, in each period the ecological question soon risks becoming instead a question of how humans relate to one another. As important as this consideration is, it is insufficient to ground a significant ecological ethic. Indeed, it suggests that the conflict between what is good for nature and what is good for poor people and poor nations will very well continue.

The contemplative dimension of a religious ecological ethic, such as proposed by Pope John Paul, may indeed extricate the tradition from this "people vs. nature" impasse. In any such system, contemplation becomes the basis for an ethics of appreciation. Looking at oneself and one's own finiteness, one discovers that there is no intrinsic reason for one's being at all. (In fact, if such an attitude is "scientific" at all, it must even contemplate the contingency and thus the possible extinction of the human race itself, while the ecosystem continues.) This creation and we in relation to creation are indeed gifts from God. Such a point of view brings with it an implicit framework from which to evaluate one's conduct in this world. A contemporary ethic, using the signs of the times, would challenge Christians to move from a point of view in which nature has little or no value apart from human choices to one which sees humanity itself as part of the larger ecosystem. Such a point of view would affect the choices which face humanity by demanding a caring presence in the world in place of a domineering one.

Contemporary ecologists suggest that the proper stance toward nature is one which avoids irreversible change, optimizes natural diversity, and emphasizes natural stability. An ethics of appreciation can enable these sorts of choices by demonstrating the need to locate oneself in and reconcile oneself to the rest of nature. The questions of justice and the common good do in fact remain, but placing these issues in dialectic with a larger ethics of appreciation provides a more adequate grounding for ethical thought, a grounding which at the same time is able to situate the human person in the world as science sees it and, because of its basic spirituality, is able to speak not only to the human head but also to the human heart.

REFERENCES

John XXIII, Pope

1963 "Pacem in terris." In Joseph Gremillion, ed., *The Gospel of Peace and Justice,* 201–41. Maryknoll, N.Y.: Orbis Books, 1976.

John Paul II, Pope

1979 "Redemptor hominis." *Origins* 8, 40 (March 22, 1979) 625–44.

1981a "Laborem exercens." *Origins* 11, 15 (September 24, 1981) 225–44.

1981b "Familiaris consortio." *Origins* 11, 28 & 29 (December 24, 1981) 437–68.

1989 "Peace with God the Creator, Peace with All of Creation." *Origins* 19, 28 (December 14, 1989) 465–68.

Leo XIII, Pope

1891 "Rerum novarum." In David M. Byers, ed., *Justice in the Market Place,* 13–41. Washington, D.C.: National Conference of Catholic Bishops, 1985.

National Council of Catholic Bishops

1991 "Renewing the Earth: An Invitation to Reflection and Action on the Environment in Light of Catholic Social Teaching." *Origins* 21, 27 (December 12, 1991) 425–32.

Paul VI, Pope

1967 "Populorum progressio." In *The Gospel of Peace and Justice,* 387–15.

1971 "Octogesima adveniens." In *The Gospel of Peace and Justice,* 485–511.

1972 "Message to the United Nations Conference on the Human Environment." *Origins* 2, 5 (June 22, 1972) 76–77.

1977 "Address on the Occasion of the Fifth World Day of the Environment." *Catholic Mind* 75 (December, 1977) 10–11.

Pius XI, Pope

1931 "Quadragesimo anno." In *Justice in the Market Place,* 44–90. Certain changes in the English translation of paragraph 81 were made to reflect the Latin. See *Acta Apostolica Sedis* 23 (1931) 204.

Second Vatican Council

1965 "Gaudium et spes." In *The Gospel of Peace and Justice,* 243–335.

4

Theological Dimensions of an Ecological Ethic

John T. Pawlikowski, O.S.M.

The Present Condition

Back in the early seventies, well before the ecological movement had achieved its current prominence, two futurists introduced us to a fundamentally new reality with which Christian theological ethics has yet adequately to grapple. Victor Ferkiss, out of the Catholic tradition, and Hans Jonas, of Jewish background, served warning that humankind had reached a new threshold in its evolutionary journey. The human community now faced a situation whose potential for destruction was equal to its possibilities for new levels of human creativity and dignity. Which path humanity would follow was a decision that lay in the hands of this generation and perhaps the next. Neither divine intervention nor the arbitrary forces of nature would determine the choice in the end. And the decision would have lasting impact, well beyond the lifespan of those who are destined to make it. It would, in fact, determine what forms of life will experience continued survival.

Ferkiss' 1974 volume, *The Future of Technological Civilization,* put the late twentieth-century challenge to humankind in these words: "Man has . . . achieved virtually godlike powers over himself, his society, and his physical environment. As a result of his scientific and technological achievements, he has the power to alter or destroy both the human race and its physical habitat" (Ferkiss 1974, 88).

Hans Jonas, speaking in 1972 to the gathering of learned societies of religion in Los Angeles, conveyed essentially the same message as Ferkiss (cf. also Jonas 1984). Ours is the very first generation to have to face the question of basic creational survival. In the past, there was no form of human destructive behavior from which nature could not recover using its inherent recuperative powers. But today, we have reached the point through technological advancement where this rule no longer obtains. Humankind now seems increasingly capable of actions which inflict terminal damage on the natural world and raise serious questions about its future capacity to sustain human life. Buckminster Fuller did not seriously exaggerate the profundity of the choice before us when he asserted that contemporary humanity stands on a threshold between utopia and oblivion.

There are those who say that theology, whether in classical or more contemporary forms, will be hard pressed to offer any serious response to the present challenge. At the very outset of the recent upsurge in ecological awareness, historian Lynn White threw down the gauntlet when he charged that the Judaeo-Christian theological tradition was the root cause of the present ecological crisis in the Western world (White 1967). The movement which arose in response to his epochal address deliberately selected a pagan symbol for Earth Day, in order to drive home the point that a break with the Christian theological legacy was a necessary first step on the path to long-term creational survival.

But critical questions about the usefulness of theology for building a sound ecological ethic have arisen within the religious community itself, and not only among outsiders. Thomas Berry, a Passionist priest, appears to hold that the Christian biblical tradition has outlived its usefulness in defining our understanding of the origins of creation. Only a willingness to adopt the modern scientific explanation of the earth's genesis will set us on the road towards ecological responsibility. I say Berry "appears" to hold such a radical viewpoint because there are some occasions in his growing library of writings on the subject which do seem to leave a continued role for the Christian theological tradition in the shaping of an adequate ecological ethic. However, in the main, Berry accords little value to the Church's theological heritage. To emphasize this point, he prefers to describe himself as a "ecologian" rather than a "theologian."

Another critical perspective comes from the process theologian Schubert Ogden. His critique is directed far more at contemporary forms of theological expression than at classical formulations. He especially highlights the deficiencies of liberation theology with respect to ecological awareness, based on its return to the "God of history" of the Hebrew Scriptures. This vision of God has little place for the destiny of creation other than humanity. For Ogden, achievement of political liberation for the human community that does not involve a concomitant liberation of the rest of creation will ultimately prove to be a pyrrhic victory (Ogden 1979). But in fairness to Ogden's overall perspective, this criticism of historically-based contemporary theology does not totally rule out a positive contribution from theology, as a whole, in the shaping of the ecological consciousness for which he calls.

In light of the assessment of the new situation facing humankind by people like Ferkiss and Jonas, and in view of the criticisms levelled at theology by many associated with the ecological movement, is there reasonable hope for using the Christian theological heritage in the development of a new creational awareness within the Churches? In my judgment the theological tradition is, finally, retrievable; in fact, its potential contribution to the emergence of an ecological ethic is indispensable. However, the retrievability will, at best, be partial, and must be accompanied by the clear acknowledgment of significant failures in the area of ecological concern within the Christian tradition, as well as a realization that the tradition, by itself, will prove insufficient to foster an ethic appropriate for the state of the twentieth-century human condition, as described by Ferkiss, Jonas, and other futurists.

Disproving the simplistic connections between the Christian theological tradition and ecological degradation in the West, posited by Lynn White and other environmentalists, is a relatively easy task, as Thomas Singer Derr has ably shown (Derr 1975). But to convey the impression that the tradition, by itself, can generate an ecological ethic appropriate for our present circumstances, as Charles Murphy does in *At Home on Earth,* is to create a false sense of security among Christians. Granted, there are sections of the tradition which have been largely ignored in the past and which, as Murphy rightly argues, can be retrieved today in the service of an ecological ethic. Yet there is considerable disagreement among scholars about their proper interpretation. This is particularly the case with regard to the writings of Thomas Aquinas. Should Thomas be seen as fundamentally an ally or a foe of environmental concern? More importantly, Murphy's approach, strongly endorsed by Cardinal Bernard Law in the introduction to the book, fails completely to deal with the claims of Ferkiss and Jonas, that humanity today finds itself in far different circumstances than all prior generations.

It is my firm conviction that only a delicate combination of retrieval and new theological construction will, in the end, yield an ecological ethic with some meaning and force. The Churches must stand willing to examine anew several central theological themes with respect to their ecological implications, if such an ethic is not to be stillborn. These themes include the vision of God, Christology, and the origins/ends of creation. In the following pages, we will briefly examine each of these themes from the standpoint of a contemporary ecological ethic.

God-Consciousness and Ecological Concern

Scholars dealing with the Hebrew Scriptures readily point out the transitions that occurred in Israel's understanding of God. At times, several important images coexisted within the community, each vying for attention. Nonetheless, though the development was generally jagged, it is possible to discern a distinct evolutionary line in the Jewish people's vision of the divine reality. The idea of a God who created and dominated nature surely was one of the first, from an historical perspective. This vision had its similarities to understandings of God found among other peoples of the age. Yet there was a notable difference, which some see as marking the beginning of the downfall of ecological sensitivity in the Judaeo-Christian tradition.

Biblical exegetes such as J. Coert Rylaarsdam have emphasized the differentiation between the biblical God and nature that is a hallmark of the early phases of the Hebrew Scriptures. Some, today, regard this differentiation in an entirely negative light from the standpoint of ecological concern. Kenneth and Michael Himes raise this point, suggesting that in denying divine status to the gods of nature in ancient mythology, who protected streams, mountains, and forests, the Hebrew Scriptures may have in fact severely lowered the status of non-human creation. This in turn opened the doors for future exploitation of these creational realms (Himes and Himes 1990, 41).

Rylaarsdam would no doubt reject the argument of the ecologists on this point, and with good reason. While it is true that the Hebrew Scriptures depict God as standing above and beyond nature, their vision does not totally divorce the divine from the natural world. On the contrary, God is definitely to be found within the natural world. To use the words of the scriptural writers, God's face can be seen within nature. Surely the many nature festivals that formed a core part of Israel's worship life over the centuries testify to the strength of the Jewish people's continuing belief that the realm of natural creation constitutes a most important point of contact with divine reality.

In the pursuit of an ecological ethic, this is definitely a point where retrieval is vital for the Churches. Otherwise, we could find ourselves falling, willingly or not, into an ethic in which natural rhythms become the determinant for ethical behavior. This was certainly the approach taken by the Nazis in their distorted and devastating effort to create a new humanity. Some have said that it is this Nazi experience that has made Cardinal Joseph Ratzinger such an outspoken critic of much of the European ecological movement. Rightly or wrongly, he is not alone in this concern. In a recent interview, French philosopher Bernard-Henri Levy has spoken in equally harsh language about dangerous tendencies in the movement:

> . . . the just concern of protecting the environment and preventing waste must be imperatively addressed. On the other hand, the ecologists' vision of the world, and the philosophy they preach in Germany, Scandinavian countries and France, have all the ingredients of a dangerous religion, as I have been warning for many years. It is in fact the embryo of a modern fascist mass movement, featuring all the symptoms of fascism, from the brutal antisemitism it nutures to the more discreet, but no less fearsome, fantasy of a good, utmost natural community and the obsession of a clean social corpus (Levy 1991, 13).

From the Hebrew Scriptures we shall also need to retrieve what biblical scholars such as W. D. Davies and Walter Brueggemann have termed the "land tradition." Brueggemann's perspectives are especially important for the construction of an ecological ethic, as well as a counter to those who look upon the biblical tradition as a root cause of the disregard of the ecosystem.

In his several pieces on the topic, but especially in his book *The Land,* Brueggemann insists that the antithesis of the God of history over against the gods of the land, so prevalent in many interpretations of the biblical materials, is outdated. Israel's God is pictured as Lord of historical events, as well as fructifier of the land. YHWH is a fertility God, who gives life, as well as an historical God, who saves and judges. YHWH is Lord of places as well as of times. Such an understanding, according to Brueggemann, puts into serious question the validity of the existentialist approach to biblical interpretation, identified with scholars such as Rudolf Bultmann. This school of thought has exercised a strong and, from the ecological viewpoint, a basically harmful influence on modern biblical understanding. Its effect among many Christians has been to circumscribe the experience of divine presence in a person's individual relationship with Christ. Little attention was, therefore, given to finding God's presence in nature.

Finally, from the Hebrew Scriptures, we need to retrieve the notion of the human community as sharing in the responsibility for the governance of creation. This dimension of human dignity has often been overlooked in Christianity. Yet it remains one of the most powerful messages of Scripture, which takes on a new relevance in light of the contemporary ecological challenge. Thankfully, the retrieval of this tradition has now begun, not merely among individual exegetes and theologians, but in official Church documents as well. Pope John Paul II in his encyclical *Laborem exercens,* as well as the U.S. and Canadian bishops' statements on economic justice, all highlight the importance of this understanding.

To appreciate fully the significance of the biblical sense of human cocreatorship we must return to the key passages in Genesis 2. Contemporary exegetes such as Claus Westermann have emphasized that, in the perspective offered in these passages, work in the form of the care and development of inherited creation is understood as a primal religious duty. The human community has been created by God and through the subsequent covenant with its Creator commissioned "to till and keep."

These verbs are to be seen as complementary, according to Westermann. "Preservation" and "enhancement" of creation, given humankind as gift by the Creator God, are coequal duties in the mind of the Genesis author. Neither can be neglected. Nor can one ever be totally sacrificed to advance the other. Maintaining the proper equilibrium will require great skill and ingenuity on the part of the human family. And it will be an equilibrium in need of periodic readjustment.

Westermann makes another important point in connection with Genesis' use of the verbs "to till" and "to keep." Though the author undoubtedly had in mind the work of the Palestinian farmer when he wrote these words, it would be a mistake to limit his meaning to the agricultural realm. This is not to downplay the importance of farming, particularly the family farm, as a way of maintaining contact with the land. But this passage from Genesis can serve as the basis for a continually expanding definition of the kind of work involved in fulfilling our biblical vocation.

If we are to retrieve properly the Genesis notion of co-creatorship, we shall have to come to a better understanding of its most controversial aspect in today's increasingly ecologically sensitive world—*dominion* over creation. The Hebrew word *radah* ("have dominion") is extremely forceful. It does not refer simply to a kind of benign maintenance over an essentially docile and pacific natural world. Its appearance in the Hebrew Scriptures is rather rare. When it does appear, it is nearly always in the context of kingship, as in 1 Kings 5:4; Psalm 72:8; and Ezekiel 34:4. Hence, it seems to have been part and parcel of the vocabulary of royal rule. And in the mindset of Genesis, royal rule was understood in an absolutist way. Here, then, we encounter the element in the Hebrew Scriptures' vision of divine likeness. In other creation myths, humanity, despite its sometimes godlike quality, remains by and large the slave and yoke-bearer of the gods. In Genesis, however, human beings are portrayed as enjoying dominion over all they survey.

It is unfortunate that the Genesis notion of "dominion" has been poorly understood by many Christians and rejected out of hand by a number of those who would wish to make environmental preservation the preeminent moral value today. There is absolutely no basis in this Genesis notion for unbridled exploitation of natural resources, which has often occurred under modern industrial and agricultural capitalism. Quite the contrary. A careful reading of the Genesis accounts of human creation clearly gives great encouragement to those ecologists who emphasize that the earth has been left in humanity's hands as a sacred trust. Human dominion over the rest of creation is not to be one of exploitation or simply dominance. Rather, it is to be a caring role modeled on God's ultimate dominion.

But neither does Genesis provide encouragement for these extreme ecological views that would leave little room for planned economic development and utilization of natural resources. And it definitely repudiates any theology of ecology that would "collapse" humanity back into the natural world in the manner of some of the nonbiblical religions. Some well-intentioned proponents of an ecologically based theology, who work out of process theology (e.g., John Cobb), or advocate Native American religion as a paradigm, in fact, threaten ecological preservation in the end by unduly diminishing the distinctive and central role the human community must now assume in directing the process of the preservation of life forms at every level. This is the point emphasized by Ferkiss, and especially Jonas. We have reached a new stage in human development, in which humanity is now being summoned to an unprecedentened exercise of its role of "dominion" over creation. To underplay this role, at the present time, is to risk even greater ecological destruction. Neither the self-preservatory powers of nature nor divine intervention will do it alone. Humankind now stands centerstage in the process.

Certainly the situation is fraught with danger. The exercise of dominion always risks the danger of misuse. In this sense, Stanley Hauerwas' warning that an emphasis on human co-creation may easily lead to the kind of haughtiness that allowed the Nazis to annihilate millions of Jews, Gypsies, Poles, gays, and others is a most important caution. The Genesis theology of co-creatorship need not inevitably result in such haughtiness if we remember that sharing in the divine creative power is a free gift of the Creator. In the face of the massive new technological and bureaucratic capacity now in humanity's reach, there clearly is need for the sense of humility underlined by Hauerwas. But, if such humility is not accompanied by the pervasive adoption of the Genesis theology of co-creatorship, redefined for our time, Hauerwas' call by itself may result in a form of human failure to take up full responsibility for the governance of the earth. Such failure would likely result in economic, ecological, and even nuclear disaster on a global scale. This point has recently been made by Lutheran theologian Philip Hefner, director of the Chicago Center for Religion and Science, when he speaks of "the continuously unfolding scientific worldview, which places our human enterprise within a vast evolutionary process and which heightens our sense of the decisive role of human agency within that process" (Hefner, 1).

One final point relative to human co-creatorship. It should be emphasized, if it is not already clear, that the understanding of this commission must reach far beyond the parameters of the biblical vision. We have an altered sense of God's relationship to humanity, made evident through such epochal events as the Nazi Holocaust. We now recognize that we must go to great lengths to recover a sense of the healing and directive power of the divine presence, while at the same time spurning any temptation to imagine that God controls the fate of creation in an immediate and direct fashion. The latter responsibility has clearly been left to humankind, a responsibility whose full dimensions are only beginning to unfold in our time.

Christology: Asset or Impediment to an Ecological Ethic

Some Christians who are concerned with ecology view Christology, at least in terms of incarnationalism, as a definite obstacle to the emergence of a new creational ethic. It is thought that the emphasis on incarnationalism within the Christian theological tradition, an emphasis still very much alive (e.g., the encyclical theology of John Paul II), has resulted in the wholesale devaluation of the dignity of the rest of creation. The divine presence is now confined, as it were, to the human realm alone, leaving all else as the virtual plaything of humanity.

That some theologians, past and present, may have given such an impression of the ultimate meaning of incarnational Christology may very well be the case. But this does not mean that it is the only possible approach. In fact, it is valid to understand incarnational Christology in a way that enhances, rather than detracts from, a basic ecological commitment. The key to this shift in perspective involves returning to the matrix of Jewish Christianity and Second Temple Judaism, where incarnationalism has its ultimate roots. The subsequent translation of the incarnational vision into exclusively Greek philosophical categories has robbed it of some of its initial significance, a significance which has become all-important in an age of critical human choice regarding the future of life forms. First of all, for Jewish Christianity, the understanding of the special, intimate relationship between humanity and the Creator God made possible through this God's unique relationship with Jesus did not exhaust the full meaning of belief. It was the core element, the central element, to be sure. But the earlier revelation of divine presence in all of nature found in the Hebrew Scriptures was not invalidated by the subsequent revelation of divine-human union in Christ. The basic thrust of the Hebrew Scriptures remained an integral part of early Christianity's belief system. It was not a mere prelude to the vision of the New Testament, but very much part of an authentic response to the challenge of Jesus' ministry.

Secondly, and even more importantly, incarnationalism was not a miraculous doctrine for Jewish Christianity, totally detached from all other aspects of faith. Rather, it was an outgrowth of a gradually evolving understanding of a new relationship between God and humanity, within the Pharisaic movement in Second Temple Judaism. This evolution took a quantum leap in the person of Jesus, one that most within the Pharisaic camp were unable to accept. Within this process, culminating in Jesus, several realities were central: (1) an increasing dignity accorded each human person; (2) a growing emphasis on humankind's ability to reshape its social environment, which in turn signalled a diminution of divine control over every important facet of human life; (3) a sense of the need for human-divine reconciliation to replace divine lordship over creation as the prevailing relational model—God was willing to endure suffering on the cross to intensify linkage with humanity. While it is true that non-human creation does not directly enter the picture, there are important constructive implications of this understanding of incarnational Christology for the shaping of an ecological ethic.

The first implication is clearly the enhanced status accorded to the individual. Incarnational Christology gives an added significance to the original Genesis proclamation of human co-creatorship. Divine power and authority now reside in the human community far more deeply and directly than earlier envisioned, as a result of God's self-sacrificing love. Secondly, this new status clearly gave the believing community the authority to reshape its social environment. By legitimate extrapolation, we can apply that today to authority over the physical environment as well. Humanity is no longer the mere captive of either divine or natural laws, either in the social sphere or the physical sphere. It has an ability and a responsibility to preserve and reshape all that surrounds it in a spirit of reverence and continuing dependency. Finally, the incarnation, once and for all, established reconciliation as the ultimate model for the divine-human-relationship. No longer will it be necessary for human leaders to fashion patriarchal models of dominance over the human majority and over the rest of creation, in supposed imitation of the "mighty hand of God." For God has shown reconciliation, not dominance, to be the essence of majesty. If reconciliation has become the primary paradigm for God's action in light of the incarnation, then it must become the primary paradigm for humanity's actions as well. And, again by extension (although Paul already alludes to this Christological implication in Romans), this reconciliation must embrace not only the divine-human encounter, but humanity's stance towards the world with which it shares the basic life process.

Apocalypticism and the Development of an Ecological Ethic

Fr. Thomas Berry, C.P., is often quoted with respect to his insistence that the development of a sense of ecological responsibility within contemporary humanity is predicated on the replacement of the biblical narrative of creation with one more consonant with present scientific understanding of the emergence of earth and its life forms. A number of biblical scholars have, on the other hand, rightly pointed out that Genesis does not mandate any particular scientific explanation of creational origins. In any case, though battles continue in some places over the teaching of evolution in schools, the controversy over earth's beginnings does not continue to exercise the same fascination over people today as does the question of earth's demise. Books predicting the so-called apolocalyptic destruction of the universe abound. They are readily available at checkout counters and other locations frequented by the general public. Berry's warning against the effect of biblical narrative would be better applied to the popular understanding that the Book of Revelation and other apocalyptic writings predict the engulfing of the world in fire and brimstone (and probably sooner than later). The threat to ecological responsibility inherent in such an outlook was dramatically illustrated several years ago when then U.S. Secretary of the Interior James Watt testified before Congress that the nation need not be unduly concerned about the preservation of its forests and other natural resources for, after all, the world was going to end soon, and catastrophically at that.

Clearly, the emergence of a comprehensive ecological ethic will involve a wholesale effort to rid public consciousness of such an outlook towards the "end of days." However the Church may eventually come to explain the final emergence of God's reign, it cannot be interpreted as involving the wholesale fiery destruction of the creation generated by God and enhanced through the divine-human partnership. As reputable biblical scholars are insisting today, fiery destruction is in fact not the actual vision of the apocalyptic texts in the Bible. Nor do they intend to present us with inside information about future history. Rather, as John Collins emphasizes, they affirm, through the use of heavy symbolism, a transcendent world, even in the midst of considerable political and cultural chaos. Far from being mired in despair about the world, the "apocalyptic imagination," as Collins terms it, provides humankind with a solid basis for a hopeful approach to a new world order. It liberates us, in the case of our discussion, from any sense of despair about the inevitability of environmental degradation. "The legacy of the apocalypses includes a powerful rhetoric for denouncing the deficiencies of this world. It also includes the conviction that the world as now constituted is not the end. Most of all it entails an appreciation of the great resource that lies in the human imagination to construct a symbolic world where the integrity of values can be maintained in the face of social and political powerlessness and even of the threat of death" (Collins 1984, 215).

The apocalyptic writings will not prove sufficient by themselves to undergird a sound ecological ethic. Collins himself says that they come down heavy on the side of divine, rather than human initiative. Moreover, their imaginative power may remain locked in the world of fantasy, rather than reality. But combined with the Genesis co-creational texts and other motivational resources, these texts can serve both to critique the effects of ecological irresponsibility and to help generate a vision of a universe where creational harmony leads to creational preservation.

Clearly we have a need for both retrieval and change with respect to the apocalyptic biblical texts. In a recent book, *From Apocalypse to Genesis: Ecology, Feminism and Christianity,* Anne Primavesi offers a constructive contrast between biblical apocalypticism and the outlook of contemporary "green apocalypticists" which illustrates well the retrieve/discard challenge facing present-day Christianity. On the one hand, the example of the "green apocalypticists" and their dire images of an ecologically spent world reaffirm the ongoing significance of these biblical apocalyptic texts, which some Christians have regarded as irrelevant for contemporary faith. Without question, they need to be retrieved. On the other hand, in showing how the green apocalypticists differ in outlook from the biblical vision, she helps us delineate some of the areas where adjustments need to be made in the scriptural approach. For example, as has already been underscored above, the direct role of the human community in ecological preservation must be understood in a far more powerful sense, even though Christians may still wish to preserve the primacy of divine initiative in a way the green apocalypticists do not. The Church will also have to adopt a much more integral view of the relationship between the present world order and the final divine reign than appears in the biblical texts themselves.

In short, dialogue with the outlook of the green apocalypticists helps Christians today in the delicate process of updating their biblical heritage while preserving it as a critique of certain dangerous trends within the ecological movement. It can help restore to eschatology an important role in concrete, constructive social change.

Final Observations

To summarize briefly the main lines of the argument advanced above, a sound ecological ethic will emerge only within a theological context where God is understood as sharing with humankind responsibility for the maintenance and development of creation, to a degree never before conceivable, and where high priority is assigned to the reconciliation of humanity with the rest of creation. Additionally, such an ecological ethic must be guided by three fundamental convictions: (1) all of creation is integral to the ongoing process of salvation, leading to the emergence of the final reign of God; (2) humankind must act in a manner that insures the preservation of creation for future generations, because the passage to the final divine reign is one of transition, not destruction; and (3) through the gift of co-creatorship men and women share with God in the process of bringing the divine reign into realization. When confronted with the massive (and ever-growing) ecological destruction that besets our planet, the temptation to question whether the Church can respond effectively in any measure can appear overwhelming. Here is where the hopeful imagination of the apocalyptic imagination, freed from its destructive misinterpretations, can serve us well. Hope can prevail over gloom. One who shares this positive vision of the Church's potential is William K. Reilly, formerly chief administrator of the United States Environmental Protection Agency. His words serve as an appropriate closure for these reflections:

> As we approach the two thousandth anniversary of the birth of Christ, the Church can be instrumental in establishing a new sense of global reciprocity, an enhanced awareness that we human beings have a duty to nurture and sustain the planet that nurtures and sustains us. And as the community of faith continues to make strong statements that integrate knowledge and faith—statements that engage the heart—I think that the confluence of religion and ecology will continue to gather force until it ultimately washes over and fundamentally changes our culture. (Reilly 1991, 25)

REFERENCES

Berry, Thomas
1981-82 "Classical Western Spirituality and the American Experience." *Cross Currents* 31:388-99.
1988 *The Dream of the Earth.* San Francisco: Sierra Club Books.
1991 *Befriending the Earth.* Mystic, Conn.: Twenty-Third Publications.

Brueggemann, Walter
1977 *The Land.* Philadelphia: Fortress.

Collins, John J.
1984 *The Apocalyptic Imagination: An Introduction to the Jewish Matrix of Christianity.* New York: Crossroad.

Derr, Thomas Sieger
1975 *Ecology and Human Need.* Philadelphia: Westminster.

Ferkiss, Victor C.
1969 *Technological Man: The Myth and the Reality.* New York: George Braziller.
1974 *The Future of Technological Civilization.* New York: George Braziller.

Hauerwas, Stanley
1982 "Work as 'Co-Creation'—A Remarkably Bad Idea." *This World* 3:89-102.

Hefner, Philip
Undated "Discerning the Times: The Challenge of Our Situation." Unpublished manuscript.

Himes, Michael J., and Kenneth R. Himes
1990 "The Sacrament of Creation: Toward an Environmental Theology." *Commonweal* 117:42–49.

Jonas, Hans
1984 *The Imperative of Responsibility.* Chicago: The University of Chicago Press.

Levy, Bernard-Henri
1991 "The Intellectual Superstar" (interview by Nathalie Siran). *The Jerusalem Post International Edition* (July 27) 13.

Murphy, Charles M.
1989 *At Home on Earth: Foundations for a Catholic Ethic of the Environment.* New York: Crossroad.

Ogden, Schubert
1979 *Faith and Freedom: Toward a Theology of Liberation.* Nashville: Abingdon.

Pawlikowski, John T., O.S.M.
1982 *Christ in the Light of the Christian-Jewish Dialogue.* New York and Ramsey: Paulist.
1986 "Participation in Economic Life." *The Bible Today* 24:363–69.

Primavesi, Anne
1991 *From Apocalypse to Genesis: Ecology, Feminism and Christianity.* Minneapolis: Fortress.

Reilly, William K.
1991 "Theology and Ecology: A Confluence of Interests." *New Theology Review* 4:15–25.

Rylaarsdam, J. Coert
1968 "The Old Testament and the New: Theocentricity, Continuity, Finality." In *The Future as the Presence of Shared Hope,* 59–83. Ed. Maryellen Muckenhirn. New York: Sheed and Ward.

White, Lynn, Jr.
1967 "The Historical Roots of Our Ecological Crisis," *Science* 155:1203-7.

5

Taming an Unruly Family Member: Ethics and the Ecological Crisis

Paul J. Wadell, C.P.

Everyone knows too much chaos brings the house down. Something like that is happening in monumental proportions to the household we call the earth. It is interesting that the word *ecology* derives from the Greek *oikos,* meaning "house" or "place in which to live" (McDonagh 1986, 17), for it reminds us that not only are we called to be "at home" on the earth, but we are also called to live "at home" with every other creature of the earth. If the earth is our home, it is surely meant for more than ourselves. The house is home to all creation and if it is not to collapse through human thoughtlessness and irresponsibility, we must learn to live with all members of the household.

But the house is under attack. The earth cannot afford the kind of human beings we have become. It cannot afford men and women who think of themselves primarily as consumers. It cannot afford people who see the natural world as something to exploit instead of reverence. It cannot afford all the waste and rubbish that flows in the wake of a lifestyle where people measure their well-being by what they own instead of who they are. It cannot afford a way of life that takes the natural world for granted and presumes that we can do to nature whatever we want without cost.

Something has to give. We are crippling the earth, making it useless for anyone who comes after us. We are not doing justice to the rest of creation; indeed, we live in a way that makes it impossible for other things to live at all. The ones who should be the most conscientious and responsible members of the household have become the profligates. Like rowdy guests, we are wreaking havoc on everything else, destroying the only home all life will ever have.

How do we learn to live responsibly with other members of the household? That is the question the environmental crisis poses for ethics. We need, in Thomas Berry's language, a "mutually enhancing relationship" not only with other human beings, but with other species of the natural world (Berry 1988, 53). We need to change our habits and practices so that we can be a benevolent presence on earth instead of a disruptive one. We need to embrace a way of life that is not so self-serving that it brings benefits to us but harm to everything else. The challenge for the human species is to learn to live in harmony with the earth community.

The world is at a crisis point. If we do not learn to think and live differently, moving from an exploitative to a respectful relationship to the natural world, the walls of the house will come tumbling down. In this chapter we will examine three strategies for altering our stance to the natural world, so that instead of being unruly members of the household, we can live justly and responsibly. First, we shall look at the importance of fear as a starting point for ethical reflection, noting its power to effect radical behavioral change. Secondly, we shall propose a model for envisioning our relationship to the natural world, one that is based not on conquest and control, but benevolence and compassion. Thirdly, we shall suggest that the expanding powers of human agency bequeathed to us by technology require a commensurate sense of responsibility for the consequences of our behavior.

Fear: Finding a Power for Drastic Change

It is best to start with the facts: the only way the earth will survive is if we make a clear, decisive break with our familiar patterns of living. Minor adjustments will not do. What is needed is something definitively new and until this is accepted the earth will be prey to deadly ways of behavior. A completely different understanding of what it means to live well is required. We are not talking about living a little more simply, but a transformation so complete that it involves a redefinition of ourselves. Thomas Berry describes it as a "psychic and social transformation" (Berry 1988, 42), a change in how we value, choose, act, and decide, and especially a change in how we understand our relationship to the earth.

For a long time many Americans have been misled by the cultural optimism of a technological-consumerist society that sanctions a way of life that can no longer be sustained without irreparable harm to the environment and injustice to future generations. If there was a time in which such an attitude was a viable philosophy of life, that has long since passed. We must let go of what we have found to our liking, and embrace a way of life that will mean less for us; however, in this case less for us is much better for everything else.

The trouble, however, is that sometimes the people who need to change most are the last ones to see it. A transformation in our cultural presumptions, perspectives, and values is required, but this kind of full-fledged change is exactly what we resist. To change would mean to abandon a way of life that is to our advantage. Many Americans have grown comfortable with affluence and resist the challenge to live with less. Few are so committed to justice or grieved by injustice as to be willing to take a step down in lifestyle. As Birch and Rasmussen observe, "Yet our long history says we will not part easily with the mind-set of conquest, control, and consumption. Nor will we easily part with a standard of living to which we have become habituated. . . . We will do all we can to stay where we are" (Birch and Rasmussen 1978, 41).

Sometimes we simply become inured to the detrimental effects of our actions. The need for change may be glaring, but it is easily missed because our cultural presuppositions shield us from the truth of our behavior. Never questioning the dominant values of our society, we lose the moral vision and sensitivity necessary to assess what our way of life does to others and to the earth.

Thomas Berry alludes to this when he speaks of the "psychic entrancement" of our technological-consumerist society (Berry 1988, 32). He is talking about the almost unassailable power of the guiding myths of a society. A myth gives the overarching interpretation of the meaning of life and our place in the world. It is through our dominant myths that we understand ourselves, gain some sense of the meaning of existence, and explain fitting or unfitting behavior. It is notoriously difficult to abandon the paradigmatic myths of a culture. They are so woven into our consciousness that it is hard to imagine new possibilities. Since the myth functions as the lens through which we view the world, we focus on what reinforces it and overlook what might challenge it. And even when the destructive dimensions of a myth become apparent, change is difficult, for it demands that we reject much of what we have come to believe about our place in the world.

Where then do we find the energy to change? It is here that fear has an important role to play in shaping an ethic responsive to the ecological crisis. Although we usually think of the "good" or the "valuable" as the appropriate starting point for ethics, it is often true that what we fear is better at getting our attention and rallying our energies. We may recognize something as good but not respond to it. But it is almost always the case that if we are genuinely frightened by something and think it will harm us, we immediately move away from it. It is easier to perceive an imminent danger than a good. If something is menacing we feel it, the threat instantly commands our attention. Our senses about harm are much sharper than our recognition of the good; thus, in order to find the energy necessary to turn away from destructive behavior, we first need to feel its evil. Put differently, it may be that only when we become aware of something as dangerous that we can recognize what is beneficial (Jonas 1984, 27).

Feelings of revulsion are important in clarifying what to avoid and what to desire. Hans Jonas, following Thomas Hobbes, argues that the starting point for morality is not love for the greatest good, but fear of the greatest evil (ibid., 28). Acute fear is our instinctive, compulsive reaction to anything which threatens the fundamental law of our nature, self-preservation, and it is in having that threat sharply defined that we more aptly grasp what will nurture and sustain life. Sometimes we recognize what is genuinely valuable only when we are starkly threatened by its opposite. In this respect, fear can bring a clearer estimation of the good. As Jonas says "the heuristics of fear is surely not the last word in the search for goodness," but it is "an extremely useful first word" (ibid., 27).

The difficulty, however, is that we can be oblivious to what we should fear the most. What makes our current situation alarming is that we may not see the danger. We can deceive ourselves about dangers, choosing to deny or overlook them, or wagering that we can postpone the consequences of our actions long enough to enjoy the benefits and escape the harms. Even fear demands enough moral sensitivity and truthfulness to recognize something as a genuine harm. Fear can be a powerful ally to behavioral change, but what happens if the danger is not accepted as truly harmful? Fear works as a source of behavioral change only when what is threatening is accepted as real, and it is by no means clear that most Americans believe the harm being done to the environment is perilous enough to prompt a change in how we live.

Still, not appreciating the destructive consequences of our behavior does not absolve us from responsibility. As moral beings we are responsible for our actions whether we intend the consequences or not. If we do not fear what we ought to fear, we have an obligation to use our intelligence and imagination to cultivate a sensitivity of the harm involved in how we live. We must nurture an awareness of the impact of our actions on the earth. If we lack the sensitivity required to recognize threats and dangers, we are living unjustly, numb to the repercussions of our actions on ourselves and everything around us.

Moreover, even if we lack a sense of danger because the effects of our behavior seem remote enough to pose no immediate threat to ourselves, we have a duty to cultivate a repugnance for the harm our behavior will bring to those who live after us. How we live today may bring immediate benefits to us, but be disastrous for future generations and the earth. We have an obligation to care about the well-being of future generations. Justice requires largesse of soul. It demands enough sensitivity to be influenced by the plight of others, even those distantly removed from us. Without the capacity to fear for them, there is no way our life today will be mindful of what we owe them tomorrow.

What does it take to make us fearful enough to start living differently? Our confidence needs to be shaken. We must be stunned into recognition, shocked into awareness of the unacceptable costs involved in how we live. The first crack in confidence will come when we realize that we can destroy the earth. The grandest illusion fostered by our technological-consumerist society is that no matter what we do the earth will always provide for our needs. We live with the illusion that the earth is "infinite, inexhaustible and ever resilient" (McDonagh 1986, 26). But the earth will not endure our foolishness. We can destroy the earth because "we live on a planet with finite, limited resources that can be irretrievably lost" (ibid.).

Let the facts frighten us. We know plant life "is absolutely dependent on a thin, fragile layer of topsoil" (ibid., 28) and that nothing can grow if this topsoil is lost, but that is exactly what is happening in disastrous proportions. Billions of tons of topsoil are being lost each year: seventy-five billion tons lost worldwide, twenty-five billion tons lost annually in Asia alone, and in Africa "land which was formerly productive is being turned into desert at an alarming rate" (ibid., 29). In general, "35% of the world's arable land is in danger of being turned into desert unless the problem is recognized and faced in a serious way" (ibid.). This wholesale destruction of the land cannot continue without dismal consequences for the earth. The earth is a fragile, delicate organism that took millions of years to form. Through our carelessness it can be lost forever.

Consider what is happening to the tropical rain forests around the world. Each year "an area the size of Cuba is being destroyed" and "by the year 2020, most tropical rain forests outside the Amazon and a small area in West Africa will have been cleared, with disastrous consequences for other life-forms on Earth, especially human beings" (ibid., 31). Why should this matter? Because rain forests are among the richest and most essential life systems in the world. The fate of the rain forest in Brazil, the Philippines, and parts of Africa is our fate too. As McDonagh notes, they "are intimately bound up with the quality of our air and water; they preserve soil; they moderate climate and affect water distribution. They are one of nature's most abundant nurseries, supplying raw material for the plants that feed and heal us" (ibid.). The tropical rain forests are home "for up to two million species of plants, insects and animals" (ibid., 35), but all these are being rapidly destroyed. McDonagh writes that "given the present rate of destruction, 25% of all life-forms on Earth could be extinguished by the first decade of the next century" (ibid.).

These facts should frighten us, but fear is not enough. Fear should lead to a change of perception; it should engender a moral vision more adequate to justice not only for ourselves, but for future generations and all members of the earth, human as well as nonhuman. We will not live differently unless we first see differently, acknowledging the relationships we have with the natural world. This is why Birch and Rasmussen say "ours is a time in which change in perception is critical to any kind of human future" (Birch and Rasmussen 1978, 57). These perceptions describe the "field of vision" through which we view the world and interpret reality. They include the core values and guiding images of our culture, our fundamental assumptions and beliefs, our most enduring attitudes and outlooks. The ecological crisis demands that our "deeply ingrained habits of mind, character, and conduct" (ibid., 72) change. This shift in perception grows from fear because fear frees us to see that our schemes of interpretation are at best inadequate and at worst deadly. Our moral vision is not enlightening but falsifying, and that is why our deeds are not just. If we let the facts frighten us we will cross over to a new way of life, ready to think and act differently.

New vision should lead to new living. The conversion in our perceptions must work its way through our whole being, ultimately resulting in a transformed self. Nothing can escape the rigors of this transformation. What is needed for the earth to survive is nothing less than a metamorphosis of the self and society, a full-scale shift in our habits of life. Fear should give birth to new moral vision; this new moral vision must be expressed in different ways of living and acting, and this widespread change of behavior should be the genesis of a better and more promising world.

What would such a conversion entail? At the very least it would involve a change in how we see ourselves. If our dominant cultural myth sees us primarily as consumers, our needs as overwhelmingly material instead of spiritual, and our satisfaction through possessions more than friends, a world view capable of justice would emphasize the kinship we have with all life and our interconnectedness to nature; more than anything, it would call us to reverence the earth instead of abusing it. If our prevailing cultural myth has encouraged an expansive, acquisitive lifestyle that says to live is always to seek more, a world view adequate to the challenges of the ecological crisis would embrace a philosophy of limits. Sometimes justice demands that we be willing to live with less even if all our wants go unfulfilled. The ecological crisis is a crisis of justice. We are living out of relationship with the earth. For the balance to be restored we must be willing to change. "The rich must live more simply that the poor may simply live" (Birch and Rasmussen 1978, 33), but of course the poor now includes the earth.

A New Way of Understanding Our Relation to the Earth

None of these changes will occur unless we rightly understand our relation to the earth. We do not have a model of the human-earth relationship adequate to the crisis we face; in fact, what makes the prevailing model most dangerous is not only that it fails to recognize the crisis, but that it exacerbates the problem. The dominant model for capturing our relationship to the earth is mechanistic. This model emerged in the sixteenth and seventeenth centuries with developments in science and technology. In her magisterial work, *The Death of Nature,* Carolyn Merchant traces the ascendance of this model to a reaction to the Black Death, widening separation and hostility between science and religion, the demise of feudalism and the rise of the commercial class, and the development of capitalism.

Unlike the earlier organic model, which saw nature as a nurturing and beneficient mother, the mechanistic model emphasized the disorder, chaos, and unpredictability of nature. This was not the nature that blessed and brought forth life, but the nature that destroyed, the nature of plagues, famines, and disasters. Nature's violence, its destructive capacities, was stressed. Not surprisingly, a threatening, menacing nature was meant to be domesticated and controlled. With the development of the mechanistic model, nature was suspect, a sleeping giant that could rise up and destroy (Merchant 1980, 192ff.).

Francis Bacon (1561–1626) is representative of the mechanistic approach to nature. He "desacralized and objectified" nature, no longer seeing "any vital inner life force from within nature" and eliminating "any lingering concern for moral rights that other members of the Earth community might have" (McDonagh 1986, 68). He did not see nature as having inherent dignity, nor as something sensitive, capable of feeling and responding; rather, nature was viewed as passive and inert. Not something to be respected, nature was to be manipulated in whatever way necessary to satisfy human needs.

Bacon wrote that nature must be "bound into service" and "made a slave" (Merchant 1980, 169). He saw nature as harboring secrets that must be wrestled from it by human ingenuity. For him, the relationship of humanity to nature was not benevolent but adversarial. Through advances in science, nature was to be attacked, ripped apart, and exploited so that its treasures could be seized and used for human well-being. The imagery is violent, but for Bacon quite fitting. To subdue nature was to render it powerless for whatever purposes human beings saw fit.

What ecologists are arguing is that if we are to survive the ecological crisis and heal the earth, we must abandon the mechanistic model and reappropriate the earlier organic model. If with the rise of modern science the "cosmos ceased to be viewed as an organism and became instead a machine,'' through the ecology movement an interest ''in the values and concepts . . . associated with the premodern organic world'' (ibid., xx) has appeared.

There are three principal characteristics of the organic model. First, it sees nature as a living organism, vital, sensitive, and responsive. Unlike the mechanistic paradigm, which views matter as lifeless, in the organic model everything in nature is permeated with life. It is because of the vitality and sensitivity of nature that human beings have to respect the earth and treat it justly. As Merchant writes, ''As long as the earth was considered to be alive and sensitive, it could be considered a breach of human ethical behavior to carry out destructive acts against it'' (ibid., 3).

Secondly, in the organic model the earth is not only a living, sensitive organism, but is also nurturing and beneficient. Like a caring mother, the earth provides for human needs. In the organic model there is harmony between nature and human beings. Nature is kind to human beings by providing them with the stuff of existence, and they are to be kind to the earth, caring for it and being grateful for the abundance of its gifts. Seeing nature as benevolent mother gives rise to a series of ethical restraints on human behavior.

A third characteristic of the organic model is that it emphasizes the interdependence of life and the subordination of individual purposes and projects to the well-being of the whole. Not only is every part of the universe connected to every other part, but ''every part of the universe and the earth was created for the benefit and support of another part'' (ibid., 23). This is a vision of a cosmos of mutual and reciprocal benevolence in which each thing exists for the sake of something else and everything lives to enrich the whole. It is not a dichotomous universe, but one of harmony and justice. Moreover, since every element of the universe contributes to the prospering of the whole, everything, no matter how small, has inherent value and merits care. Nothing is insignificant or superfluous. In the organic model the universe is utterly alive, a vast, complex organism of interconnected parts in which each part affects and influences every other part.

If the organic metaphor were retrieved, what might be some of its implications for ethics? First, it would entail a broader norm for assessing the morality of human action. If we are intimately connected to the earth, we need to consider not only what is best for human beings, but also what is best for the earth community of which we are a part. To determine morality solely in reference to human nature is unjust because we have relations to creatures other than ourselves. Moral assessment must take into account ''the needs of the entire human-nature system'' (ibid., 95).

If we are called to be accountable for the well-being of creation, the scope of moral assessment changes, shifting from anthropocentrism to biocentrism. A responsible choice is one that brings life not only to ourselves, but to the whole of creation entrusted to us by God. Put differently, we are summoned to choose life not only for ourselves, but for all creation; our responsibility is that extensive. A good moral act is one that maximizes life for the whole of nature; by contrast, any action that might be immediately beneficial for us but harmful for creation as a whole would be immoral. As Merchant puts it, "Environmental ethics is thus not merely an ethic *about* the environment but an ethic determined by it as well" (ibid., 96).

Secondly, the organic model would enrich our understanding of stewardship. Stewardship is not a license to do whatever we want with creation; rather, it describes the special responsibility we have as "the heart and mind of the universe" (McDonagh 1986, 88). We are the trustees of creation, not its lords. From the perspective of stewardship, what distinguishes us is not so much our rank in creation, but our special responsibility "to make decisions that will bring life and blessing into the created order" (Birch and Rasmussen 1978, 106). To be stewards of creation does not mean that we stand apart from the earth, but that we are accountable for how we act towards the earth. Creation is God's, not ours, and God has called us to be responsible for an earth that is an expression of God's love. We are called to befriend the earth so that it brings glory to God, not sadness.

Finally, the organic model broadens our understanding of who is our neighbor. It extends the category of neighbor beyond those close at hand to include all who will come after us. Since future generations will be impacted by how we live now, we have a responsibility to take their welfare into account. Birch and Rasmussen call this our "intragenerational responsibility" (ibid., 181).

But just as importantly, the organic model challenges us to consider nonhuman nature our neighbor too. If we are connected to the whole of creation, justice demands that we consider the repercussions of our behavior on nonhuman nature. Our neighbor is all of life, and just as we are called to care for and be responsible to other human beings, we are called to care for the earth. "Our anthropocentric perception needs to shift to biocentric perception, justice for the full community of life" (ibid., 182).

We have to draw the lines of community more broadly. We think community includes only other human beings, creatures like ourself; consequently, we conclude that justice extends no further than to those like us. But our sense of accountability grows when the boundaries of community are extended. Community signifies all those we are willing to include, and thus for whom to be accountable. There may be degrees of accountability, but we cannot be indifferent to anyone who falls within the boundaries of the community. Shifting to the organic model reminds us that the earth community includes nonhuman life too. We must recognize the kinship we have with the natural world and realize we cannot disregard its well-being. Like the Buddhist, with the virtue of compassion, we are called to show tenderness of heart to all reality, cultivating sensitivity for all things living.

Why New Powers Mean New Responsibilities

Ethics is about actions, about ordering and regulating our power to act so that it remains a constructive rather than a destructive force. One of the reasons this "house" we call planet earth is tumbling down is that our ethics has not kept pace with our powers to act. As Hans Jonas suggests, "novel powers to act require novel ethical rules and perhaps even a new ethics" (Jonas 1984, 23). We need a new ethics because technology has given us new powers. We are trapped in a dilemma because we employ an understanding of ethics whose principles and parameters are outmoded. Most models to contemporary ethics focus on individual acts and individual responsibility; they give primary, if not total, attention to interpersonal relationships, but little attention to the earth-human relationship.

We have inherited an understanding of ethics ill suited for the ecological crisis. Our ethical systems were designed to address a much more limited conception of human power and agency. They work well when the range of human activity is narrow and easily defined, and when its consequences are highly predictable, but break down when the effects of an action include not only interpersonal relations but future generations and the earth as well.

We are in an entirely new situation, which is why we need a new understanding of ethics and responsibility. Because the nature of human action has changed, ethics too must change. The problem is that our moral thinking has not kept pace with our power to act. The difference in the scope and power of our actions today is such that there is little in conventional ethics to help us. This is behind Jonas' remark that "the qualitatively novel nature of certain of our actions has opened up a whole new dimension of ethical relevance for which there is no precedent in the standards and canons of traditional ethics" (ibid., 1).

Advances in technology have vastly extended the power of human agency. Prior to the development of modern technology, human ingenuity was used to order nature and make it serve our needs, but there was an obvious limit to what could be done. Compared to the greatness of nature, human power seemed puny. Though we could shape nature, the earth was largely untroubled by our intrusions. Because of this, what we might do to nature was not considered a moral question but a technical one. Ethics applied to human relationships in which what one person did to another could make a difference, but that hardly seemed the case with our actions towards the earth; consequently, dealings with the nonhuman world were judged to be ethically neutral.

Similarly, the ethical appraisal of the consequences of actions did not extend beyond the proximate reach of the action. What mattered in ethical analysis were immediate consequences, not more remote ones. Therefore, the more distant repercussions of an act, whether on future generations or the earth, were not judged to be part of the action, nor the agent's responsibility.

But everything has changed. If technology has redefined our powers of acting, should not our understanding of responsibility be similarly extended? A general principle for ethics is that we are responsible for the well-being of whatever our behavior can significantly affect. If before we were not held accountable for the earth because our actions were not believed to have lasting impact on nature, now we are responsible for the "whole biosphere of the planet . . . because of our power over it" (ibid., 7). Our responsibility is commensurate with our power, and because our actions today have a bearing on much more than human interest alone, we have a duty to inquire about the impact of our actions on nonhuman life. The earth has a moral claim on us because it can be damaged, even destroyed, by how we live.

Furthermore, thanks to technology, our actions can have consequences that are not only of unimaginable magnitude and irreversibility, but that are also cumulative. Cumulative consequences are those in which one effect builds upon another; they grow in magnitude the further removed they are from the original action. Distance from the original act augments instead of weakens them. In a technological age much of what we do has cumulative effects. Our acts may appear benign to us because of their immediately good results, but they can do harm if their consequences, instead of diminishing, gain momentum and have effects totally contrary to what we intended. With cumulative consequences, the benefits of an action are often disproportionate to its ultimate costs.

That is why caution and restraint must become key elements to responsibility. Jonas expresses this by saying that when we are unsure about the long-term consequences of our actions, especially possibly negative ones, we must give the benefit of the doubt not to the "prophecy of bliss" but to the "prophecy of doom" (ibid., 31). In an ecological age, responsibility means no action which would bring benefits to the present but jeopardize the interests, needs, and well-being of future generations could be justified.

Moreover, this emphasis on caution suggests that instead of focusing primarily on possible good consequences of an act, our attention needs to be drawn to potentially negative results. In our ethical calculations, greater place must be given to what is possible, especially to projections of harmful consequences. Restraint and tentativeness should be hallmarks of our behavior. This is important because of our tendency to dismiss any predictions that might make us question a proposed action from which we might benefit, but others may suffer.

Confronted with the ecological crisis, responsibility also demands that we recognize what we do not know as much as we acknowledge what we do. We need to know what the effects of our behavior might be, but developments in technology make it much more difficult to be certain of what an action involves. If in the past a kind of everyday prudence was sufficient for guiding us through the thickets of life, that is no longer the case. Our increased powers require supreme wisdom, but such wisdom may not be available to us. Our situation calls for a new kind of humility, a humility which realizes our power to act may far exceed the wisdom we have to act responsibly. As Jonas says, our "predictive knowledge falls behind the technical knowledge that nourishes our power to act" (ibid., 8). That is ethically important and suggests we should not always do what we have the power to do. We should postpone what might be technically possible until we have acquired sufficient knowledge of the consequences. Where our power to act can effect the condition of life across the globe and the possibility of life for future generations, we have a duty to recognize our ignorance, and to allow that confession of uncertainty to enter our ethical analysis. In short, if we lack the predictive skill to see clearly the ultimate implications of our behavior, we should embrace an ethic of responsible restraint.

In this chapter we have examined some of the challenges the ecological crisis poses for our understanding of morality, giving special attention to three strategies for altering our relationship to nature. We looked at fear as a starting point for ethical reflection and a power for working radical behavioral change. We examined a model of nature that would enable us to live justly towards the earth instead of ravaging it. And we saw why our increased powers of action demand a different sense of responsibility, especially a greater appreciation for the repercussions of technology.

What all this suggests is that something has to give. The earth is in a crisis it cannot long endure. Whether the planet has a future hinges on our willingness to change. The earth cannot afford us as we are. We are the unruly members of the household who need to be tamed. If we are to live at peace with the rest of creation, we must become people of benevolence and justice. If we do the household will flourish, but if we refuse the chaos will continue and the house will come tumbling down.

REFERENCES

Berry, Thomas
1988 *The Dream of the Earth.* San Francisco: Sierra Club Books.

Birch, Bruce C., and Larry L. Rasmussen
1978 *The Predicament of the Prosperous.* Philadelphia: The Westminster Press.

Jonas, Hans
1984 *The Imperative of Responsibility.* Chicago: The University of Chicago Press.

McDonagh, Sean
1986 *To Care for the Earth: A Call to a New Theology.* Santa Fe: Bear & Company.

Merchant, Carolyn
1980 *The Death of Nature: Women, Ecology and the Scientific Revolution.* San Francisco: Harper and Row, Publishers.

Part III

Liturgical Perspectives

6

Liturgy at the Heart of Creation: Towards an Ecological Consciousness in Prayer

Richard N. Fragomeni

In May 1984, the International Committee on English in the Liturgy (ICEL) published its first original Eucharistic Prayer text composed in English. After a year of consultation with bishops and liturgical experts throughout the English-speaking world, the text was revised at the end of 1985. The revised text was entitled "Eucharistic Prayer A," and was published by ICEL in January 1986 (ICEL 1986).

Of the eleven conferences of bishops that constitute ICEL, ten accepted the prayer text, while the United States conference voted it down at its November meeting in 1987. At this annual gathering, the bishops were persuaded that the text raised questions of doctrinal fidelity. It seemed too poetic and too imbued with evolutionary images of creation. As a result of the vote of the United States bishops, the prayer was denied Roman approval for all the other conferences that had accepted the text. It now lies dormant, mostly unknown.

What is interesting about this prayer, ironically, was rightfully noted by the United States bishops. The prayer makes a vital attempt to incorporate language of creation into the euchology, underscoring "the gift and responsibility of human beings as 'creatures having language,' as 'poets of creation' whose speech gives voice to all living things in praise of God" (ICEL 1984). The appropriation of the Genesis myth in the text is done in such a way that acknowledges the evolutionary theories of creation and the stewardship understanding of humanity's role in creation. In other words, it affirms contemporary theories of science and transposes the role of the human in creation from one of dominance and control to one of stewardship. In this regard, the text attempts a language of poetry enlightened by a second naivete, that is, a scientifically responsible interpretation of the Genesis myth which is, nevertheless, appreciative of fundamental symbols and their importance in culture and consciousness.

For instance, the preface section of the prayer suggests the evolutionary character of the world's history and links it to creation in these words:

> In the beginning your Word summoned light:
> night withdrew, and creation dawned.
> As ages passed unseen,
> waters gathered on the face of the earth
> and life appeared.
>
> When the times at last had ripened
> and the earth grown full in abundance,
> you created in your image man and woman,
> the crown of all creation.

Thus, this text can be interpreted as an attempt to unite a contemporary scientific perspective of creation and the metaphoric and mythic portrayal of the origin of the world in the Genesis account.

Although the prayer was denied ecclesiastical approbation, and by some considered a failed attempt at creativity, it nevertheless can be regarded as a heuristic text that raises several important questions for those concerned with liturgy and with matters ecological.

In its attempts to incorporate images of creation and re-creation, the text raises the question of how in a post-modern scientific era we speak in public worship about creation. Can we naively use the Genesis myth any longer? Next, it alerts us to the need of finding rich languages that can both bridge the scientific and the metaphoric, and yield powerful images to human consciousness that will trigger a response of gratitude for creation and a profound unity of respect in the human sphere of ecological responsibility. If liturgy is to speak credibly of creation and of the graciousness of nature, what language and images does it employ?

Urged on by the questions raised by Eucharistic Prayer A, this essay will make several points. First, it will examine the issue and implications of speaking of nature and creation in a religious context when a naive use of Genesis is no longer appropriate. To assist this discussion, a text by Walter J. Ong, S.J., will be reviewed. As an example of a creation poetics, sensitive to the exigencies of contemporary science, the work of Brian Swimme and Thomas Berry will then be discussed. This work may provide a heuristic model and creative insights for the incorporation of a variety of languages of cosmic wonder and ecological sensitivity into liturgical prayer. Finally, the essay will explore the possibilities of these insights for Christian liturgy. How could such insights enable Christian liturgical assemblies to be celebrations which trigger a deep awe and gratitude for nature and a singular appreciation for the stewardship of creation? In other words, in what ways can a renewed narrative of creation be incorporated into a Christian liturgical assembly so as to give expression to the stupendous depths of creation and engender praise?

I. Speaking of Creation in a Scientific Age

A. The Problematic Identified

In an article entitled "God's Known Universe and Christian Faith: Pastoral, Homiletical and Devotional Reflections" (1991), Walter J. Ong, S.J., addresses the issue of whether or not Christians, in the life of faith, have been conscious enough of the depths of the real universe, as they have been articulated in contemporary science. He argues that if the Christian tradition is to remain faithful to its own faith it must vastly enlarge its perspectives concerning creation and the world of nature in which we come into contact (Ong 1991, 255).

Grounding his thinking in an evolutionary understanding of creation, Ong sounds his warning clearly:

> When we think today of God's creation in a pastoral, homiletic, devotional, or even formally theological setting, it appears that we do not normally think of the entire, real universe that we now know God created. This universe is now known to be some 15 billion years old, more or less. It consists of billions of galaxies. It is a universe in which to the best of our knowledge our own human species, *Homo sapiens,* has been present for tens or even hundreds of thousands of years. (Ibid., 243)

Ong considers that scientific understandings about the vastness of the universe and the fuller reality of creation, should give the Christian believer an "even more immediately urgent sense of God's majesty." This appreciation of science as a portal for awe seems to indicate clearly the need for a critical assessment of the scientific as a mode of understanding reality. In some sectors of culture the vastness of scientific experimentation and discovery, rather than instilling a sense of wonder at creation, can lead to the hubris of thinking that by science and scientific discovery, nature and creation can be controlled. In this case, the control seemingly established by scientific discovery has led to a disrespect of nature, and the fatality of ecological abuse and crisis. The question then arises as to what kind of language, informed by scientific data, will generate awe and not hubris. Certainly, the environmental crisis we now find ourselves in raises the question of how we dispose of nature at will, thinking that we have control over it by our science. The Christian appropriation of the scientific must bring about a greater appreciation of creation.

Ong, therefore, alerts us to the fact that Christians have much more to think about and wonder over than those who inhabited biblical times. He calls for a renewed liturgical, devotional, and theological rethinking of creation. This rethinking will necessarily have impact on the religious imaginations of Christian believers and on their relationships to each other and the world. Ong sums it up this way:

> In our liturgical life or other devotional life, in our struggle to face the world's moral problems, so many of them brand new, in explaining the faith to ourselves and others, in working with others for faith and justice and other related ends, the temporal and spatial model of the universe that lies at the base of consciousness makes a difference. Our growing sensitivity to ecological problems, which are human problems touching our relations to the real universe and thereby to our real fellow human beings, desperately needs theology—systematic, moral, pastoral, ecumenical—in an up-to-date cosmological setting. In today's information culture, everything is becoming more and more explicitly related to everything else at always accelerated speed, and we need an actively up-to-date and deep cosmological sense to be able to imagine and adjust ourselves to what is going on and to bring Christian faith into better contact with what is going on. (Ibid., 246)

Ong is convinced that the scope of the fullness of the universe has not yet penetrated the collective consciousness of Christians. This perspective must become a new habit of consciousness if we are to render to God full praise and gratitude for creation, for the incarnation, for life itself. In view of this present essay, Ong's concern is germane. Can the liturgy and the language of liturgy incorporate the findings of scientific studies in such a way that the consciousness of Christian believers would be converted into a new creation sensitivity and into an interdependence with nature?

In the context of such a renewed understanding of nature and creation, which he claims is imperative in a renewed theological, liturgical, and pastoral consciousness, Ong continues his discussion by offering five "special problems raised by the well-known, larger and more realistic view of creation." These problems elaborate the dimensions of understanding that a renewed Christian practice must acknowledge.

The first, Ong claims, is that persons in Western culture live in narrowly constructed psychological worlds, concerned with personal questions of existence and meaning. The challenge is to bring these personal microcosms into communication with the wider macrocosm of the actual world of human relationships and natural networks of ecological dependence. Persons are to be awakened to the transformed nature of the world, reshaped by mass communication and scientific awareness. In view of this broader understanding of the personal world, religious thinking must awaken a deeper appreciation of the "immensity and dignity of all human communities and of all creation, past and present, attended to more explicitly than we have generally attended to such matters thus far'' (ibid., 250).

Secondly, Ong considers the greatest challenge to be the opening of consciousness on the level of culture and Christian communities to the ''real dimensions and implications of the universe which God has created and in which the Son of God became man and redeemed us'' (ibid.) He calls for a conversation between scientists and theologian/preachers, so that the universe, explored by science, can be given authentic speech as God's universe. Without a communal shift in understanding the universe the significant questions of ecology will not be sufficiently addressed.

The third problematic brings Ong to the writings of the founder of the Jesuit community, Ignatius of Loyola. Ong acknowledges that his discussion of creation finds its roots in the ''First Principle and Foundation,'' elaborated at the beginning of the first week of the Ignatian *Spiritual Exercises.* Understanding the notion of ''foundation'' as an invitation for a spiritual conversation with and about creation, Ong suggests that a personal and communal relationship with God, informed by a larger cosmology, will raise new ethical questions that must be addressed. It is to this conversation that Ignatius invites those who follow the course of the *Exercises.* In view of the broader understanding of creation Ong states: ''The matters that geneticists are now involved with entail countless personal moral questions and questions bearing on our understanding of our own humanity that could well enter into 'spiritual conversation.' Such conversation could for many persons involve momentous ecological problems with which they are personally involved, often mingled with ethical problems such as those of financial profiteering and various forms of oppression of the poor'' (ibid., 252). The consciousness of the expanded universe, therefore, is one that invites an engagement of thought and the possibilities of renewed understandings of responsible living in the context of creation.

The fourth issue which must be attended to is that of the interpretation of ancient biblical texts in view of the present understanding of creation. Ong admits that a total recapturing of the past is impossible; yet it can be recovered into the present by means of interpretation. Here it seems we are at the heart of the issue. What is the language that will allow us to interpret the past biblical and theological tradition within the contemporary understanding of creation? For Ong, this becomes a significant task for the Christian Church.

The final direction that Ong takes is to argue that perhaps the people of faith are not prepared for a new kind of ''hypercosmological thinking'' and the types of questions which it raises for people who profess a belief in God (ibid., 254). Ong warns of such hesitancy especially when it manifests itself as an indifference to global and futurist questions to which Christians are necessarily committed by virtue of their profession of faith in the incarnation.

In summary, Ong's article suggests that the task of liturgy along with homiletics, theology, and devotional practices is to bring "the real world as we now (still imperfectly) know God created it into our Christian life." This, he claims is "radical to the Church's survival and hence in God's loving providence can and will be accomplished" (ibid., 257).

B. Towards a Liturgical Language of Creation

From this discussion we can identify several essential insights which Ong's text makes pertinent for the issue of the language of creation in the liturgy.

First, to be faithful within the Christian tradition, an ongoing incorporation of images of nature and creation as the locus of God's activity must enliven the religious imagination of believers. This incorporation will allow the full range of human consciousness to be caught up into the grandeur and wonder of the created order in which humanity exists and is redeemed. Without fresh images of creation, grounded within the discoveries of contemporary science, the person of faith will inhabit a shallow reality.

Second, the framing of these images will be a collaborative task between poets/theologians and scientists. The conversation between science and religion is a necessary channel for both honesty in imagery and integrity of a faith tradition. Religious appropriation of scientific data would seek to foster a feeling of awe even over such things as the size of the universe. This response could transform the all too common attitude of hubris in the face of nature, pride in the human ability to accumulate data. This attitude of dominance has led us to the present environmental crisis wherein we seek to dispose of nature by controlling it. Thus, scientists and religious persons would work together to establish informed images that trigger, as it were, an ecological sensitivity and reverence in face of the vastness of nature.

Third, the premier locus where such images and language would be proclaimed in the Christian context would be liturgical assemblies. Languages and images informed by science and grounded in metaphors of creation, faithful to the tradition of Genesis, would pervade the entire liturgy. The scientific-religious collaboration would be a consistent creative ground for the utterances of the liturgy. To craft these languages, modes of interpretation would be employed which would free the ancient texts and images to bear richer meanings and awaken many to a new mode of being-in-creation.

Fourth, along with the wonder of creation, brought forward in the Genesis tradition, a renewed language of creation must also be able to speak the wiles of evil and destruction. The resistance of many to receive the macrocosmic understanding of reality, preferring instead an egocentric use of nature's resource, is one among many indications of the need for redemption. A liturgical language, ready to utter the wonder of creation, must also be able to speak the reconciliation needed between the limited horizons of humanity and the vastness of God's creation.

These insights prompt the next question for our consideration: are there places in contemporary literature where the dialogue between science and a religious tradition has yielded a form of language that might assist the formulation of an enlightened liturgical poetics? To answer this question we move to the work of Thomas Berry and Brian Swimme.

C. A Model of Creation Narrative for Liturgical Consideration

Thomas Berry, C.P., is an historian of cultures whose work in ecology is well known: he is concerned with the survival of the human presence on the earth (Berry 1988). In an article which outlines Berry's contribution to the formulation of a language of creation which can inspire a responsible awe in its hearers, Brian Swimme, a mathematical cosmologist, and now a collaborator with Berry, articulates the main features of Berry's cosmological vision and the narrative of creation which it yields (Swimme 1987, 85). These seven features form the infrastructure of Berry's thought, from which a poetic language and narrative of creation are emerging. Swimme claims to offer only a dim picture of a vast vision offered by Berry. What follows is a summary of Swimme's observations (Swimme 1987, 85–89).

1. *The Foundations of Berry's Cosmological Vision.* First, for Berry, the greatest achievement of the present scientific era is a cosmic creation story. Familiar with the cultural role of myths of origins and their powerful significance for individuals, Berry has noted that the significance of modern science is its eventuating a new creation mythology that can deepen our realization of our bondedness and communion with the living and nonliving universe. Thus, secondly, the new creation story that is emerging is based upon empirical data. This has significant ramifications. Berry invites us to consider the fact that in the past, various creation narratives accentuated differences between members of various cultures and religions. In a transcultural creation story that emerges from a common use of science, there is now the possibility of a convincing story that could inspire a common reverence and set of values of life and nature. This feature of Berry's work has been criticized by those who question the possibility of a transcultural story in the face of the multicultural dimensions

of humanity and human life. If these criticisms are true, what can rather be offered is a common cultural commitment to the paradigm of science, which would call for a variety of creation narratives.

In the third place, Swimme notes that Berry's work emerges from a time-developmental understanding of the universe. That is, Berry claims that the dynamic of time reveals itself as an ongoing creativity. Without such an understanding of the universe, Berry considers all human thinking to be doomed to conceptual frameworks in the midst of collapse.

Fourth, Berry asserts that everything in the universe is genetically related. He acknowledges that all things, living or not, are offspring of a common supernova explosion. It follows that the universe is integral with itself. This fifth feature implies that "every being on earth is implicated in the functioning of the earth as a whole; and the earth as a whole is intrinsic to the functioning of any particular life system" (Swimme 1987, 88). In this perspective, life is a vital principle, intrinsic to the structures of the universe itself.

Berry's sixth foundation is that humanity is a celebratory species. In other words, in humanity the universe has joyfully become conscious of itself. Needless to say, one may critique Berry on this point by acknowledging the possibility that in humanity the earth may find not only its voice of celebration but also that of destruction.

Lastly, Berry posits three basic laws of the universe: (1) The universe is a differentiated reality. It is not homogeneous; everything comes to us as a unique reality. (2) The universe consists of acting realities. As we go more deeply into matter, we realize that there is nothing inert. In this, Berry claims that the universe is made up of subjects. (3) The universe is a communion of interbondedness. By gravity, electromagnetic interaction, and genetic information all is interwoven. Humanity lives in this interwovenness.

It is within this framework, put into systematic form by Swimme, that Berry operates. Thus, it provides Berry and Swimme a context out of which they have recently collaborated to craft a story of the universe. We now turn to this effort and examine its potential for a renewed liturgical appropriation of creation.

2. *The Universe Story: Narrative of Creation.* In a coauthored volume entitled *The Universe Story: From the Primordial Flaring Forth to the Ecozoic Era—Celebration of the Unfolding of the Cosmos,* Swimme and Berry have constructed a story that has as its primary referent "the great story taking place throughout the universe" (Swimme and Berry 1992, 5). The authors admit that the creative dimensions of the universe's story are "too subtle, too overwhelming, and too mysterious ever to be captured definitively" (ibid.) Thus, they proceed to develop a narrative that hopes to awaken a rich human and celebrative participation in the

universe's story. They claim that a comprehensive story of origins is needed in a contemporary era that has lost its rootedness and wonder in the earth. A story of origins offers meaning to existence, gives guidance to human affairs, is the foundational referent for personal and communal conduct and establishes the ordering of social authority. It is the goal of such a narrative to awaken an entrancement with existence. Without such entrancement, Berry and Swimme believe that "the human community will not have the psychic energy needed for the renewal of the Earth" (ibid., 268). In this regard they offer a telling comment:

> This entrancement comes from the immediate communion of the human with the natural world, a capacity to appreciate the ultimate subjectivity and spontaneities within every form of natural being. We are discovering anew our human capacity for entering into the larger community of life, something that we have not experienced in any adequate manner since our Neolithic origins. This new experience enables us to activate the more extensive dimensions of our own being. Indeed our individual being apart from the wider community of being is emptiness. Our individual self finds its most complete realization within our family self, our community self, our species self, our earthly self, and eventually our universe self. (Ibid.)

That which can trigger this entrancement and its unfolding of communion in the universe is the narrative and ritual events that awaken in human consciousness the profoundness of the story of creation. In this hope the above text is written.

The ritual elements of the narration of cosmic stories is attended to in this text. The authors are well aware that the telling of the stories of creation were accompanied by drumbeats, songs, chants, and dances which expressed the communion with reality the primal peoples appreciated and celebrated. The music and ritual of body and dance were primal responses to the depths of cosmic music which pulsed within human depths thanks to their awesome appreciation of the universe.

This primal appreciation of the awe of the universe is now to be discovered in a second level of awareness, one informed by the insights of science. In describing the ways in which earlier peoples watched the Milky Way a new awe can be retrieved by a scientific understanding of creation.

> Though we may be as dedicated to the wild spirit of the night sky, no eye of the universe will appear from behind a cloud. Nor do we need such an experience to realize what that ancient hunter came to realize. For after such long centuries of inquiry, we find that the universe developed over fifteen billion years, and that the eye that searches the Milky Way galaxy is itself an eye shaped by the Milky Way. The mind that searches for contact with the Milky Way is the very mind of the Milky Way galaxy in search of its inner depths. (Ibid., 45)

The book poetically weaves together the cosmic story beginning with the great flash and continuing through the appearance of the galaxies, the supernovas, the sun, the living earth, and the appearance of plants, animals, humans, and nations. The tale comes to its celebrative conclusion as the authors portray what they call the Ecozoic era.

Employing traditional scientific categories that divide the eras of earth development, the Ecozoic era is characterized at its most fundamental level as a time of mutually enhancing relationships between humans, the earth, and creation. In this era the universe is appreciated as a communion of subjects, rather than an assortment of objects (ibid., 243). The ultimate goal of this new era as it dawns in our consciousness is a commitment towards human consensus, and a movement away from the devastation of creation by the arrogance of science, misused in a forgetfulness of creation. "What the Ecozoic era seeks ultimately is to bring the human activities on the Earth in alignment with the other forces functioning throughout the planet so that a creative balance will be achieved" (ibid., 261).

This creative balance demands an interspecies well-being, manifested in forms of governance, education, religion, and ethical norms (ibid., 260). The Ecozoic era is a time of integration with the energies of the universe itself. It is for the sake of such an integration that Swimme and Berry have narrated their cosmogenic narrative, steeped in a scientific paradigm and written to expand the consciousness of the reader.

D. Summary Reflections

This narrative endeavor of Swimme and Berry allows us to make several reflections about the telling of the story of cosmic origins.

First, the narrative of origins must be faithful to the present understandings of the universe that are available by means of scientific observation and experimentation.

Second, the language of this narrative must itself inspire awe in the face of the larger cosmic story which it seeks to mediate. It is a language of poetry and narrative that finds its grounding in the body and senses. The telling of the narrative is completed by drums, dance, and the pulse of energy which is the energy of the universe itself.

Third, various cultures and modes of expression will complete and enhance the universe story.

Fourth, in an age of destruction of the earth, the need for a new story of creation is an imperative.

Fifth, the telling of this story in its various modalities serves an era of awakening communion with all creation. This communion is made conscious in humans who can celebrate the wonder and awe-inspiring depths of the mystery of the universe.

Sixth, a serious question arises from the work of Berry and Swimme. This question is raised here simply for future discussion concerning the pertinence of this work for a liturgical appropriation. Do Berry and Swimmer (and we may include Ong as well) wish to substitute a scientific narrative for a myth? Why tell a creation story or sing a creation poem at all? It seems that the genre of myth has little to do with a *fact* of creation, or with any time sequence, but with how humanity lives in this universe, relates it and itself to various life-forces, and finds the promise of God abiding poetically therein. Regardless of this issue, yet to be explored, the insights and invitations of these writers can be rich fare for our next consideration.

In view of the imperatives set forward by Ong and the narrative language created by Swimme and Berry, what we have are strands of insight which can direct a renewal of liturgical language and practice in an age of renewed cosmic awareness. What Berry and Swimme offer us is an example of how we can speak of nature and creation when a naive use of Genesis is no longer possible. The invitation is, therefore, to create narrative and ritual modes, sensitive to the multi-cultural dimensions of the human community, which will create astonishment anew over the vastness of creation.

With these summary insights, we come to the second question of this essay: how and where might we use the insights and poetry of this renewed cosmic narrative within the Christian liturgical assembly? It is to this discussion that we now turn our attention.

II. The Narrative of Creation and Christian Liturgy

While it may be premature, and, thus, impossible to offer definitive ways in which the language of a renewed cosmic story may be integrated into Christian liturgy, this section endeavors to point to fields within the present liturgical life of the Christian community which may employ the insights of this conversation. There are two: the liturgical calendar, and the field of body and symbolic manifestation.

A. The Calendar of the Liturgy

In a telephone conversation with Thomas Berry, I was struck by his interest in the liturgy and the liturgical calendar. There is, no doubt, a resonance between his thought and that of Ong, that the liturgy must attend to the new understandings of the universe as a way of penetrating the religious imaginations of believers. In this particular conversation, Berry suggested that the liturgical calendar should include celebrations of cosmic significance. Quite in tune with other "doctrinal" feasts, such as Trinity Sunday, Berry proposed feasts which celebrate creation, such as a feast of the Lighting of the Sun, or a feast of the Creation of the Universe.

While I am not convinced of the propriety of celebrating particular feasts of this order, Berry's idea may provide food for thought. It raises this important question: in the present liturgical calendar do feasts and seasons already exist which could be given a stronger cosmological orientation in their celebration? An examination of the present Roman liturgical calendar, indeed, offers the possibility of retrieving the cosmological symbolism of the great feasts, especially in the celebration of Pasch and Christmas.

1. *The Celebration of the Pasch.* The present celebration of the paschal night already includes a proclamation of the story of creation from the first chapter of Genesis. This narrative of the days of creation concludes with the proclamation that God rested, after seeing that everything was very good. In an early Church tradition, however, evidenced in the Armenian Lectionary, the proclamation of creation used to include the story of sin, the basis for humanity's travail on the earth. In view of this early tradition, could not the story of creation, proclaimed at the vigil, include the narrative of sin as a pointer to the ecological problem of today, that is, to the hubris of humanity in the face of creation, and a reminder of the need for the cultivation of gratitude and grace in the promise of Christ?

To give the creation story at the vigil a riper symbolism, in keeping with contemporary observation of the universe, it could be complemented by a version of the Canticle of the Three Children, from the Prophet Daniel. Not only was this canticle a traditional component of the vigil, but it also includes a praise for the galaxy, giving a voice to all creation in praise of the covenant that abides in spite of sin. The proclamation of the paschal night would thus incorporate an incantation of praise in the mouth of creation, and in the redeemed mouths of humanity.

Along with the creation narrative, the vigil also includes prophetic texts which speak of the expectation of the new heavens and the new earth. These texts, as part of the covenant of God's promise, echo loudly the significance of creation and the longings of a people of faith. In the promise of the resurrection, a new commitment to live in this universe and relate to its life-forces with the redeemed energies of compassion and gentle abiding is uttered on the paschal night.

The paschal vigil is a night vigil, in which Christians await daybreak and the bright promise of the morning star. A strong sense of this expectation could be evoked in the blessing prayers, which are a significant part of the vigil. Attention could be given to the blessing of fire, the blessing of light (the paschal candle), and the blessing of water. Blessing prayers serve as poetic incantations, and can well be servants of the religious imagination and the dynamics of conversion.

Presently, the Roman Sacramentary has only a short oration for the blessing of fire, in which no cosmic or creation imagery is used. The development of a blessing prayer over the paschal fire could include powerful evocations of the sun and stars as well as the praise of God in the powers of life and death known in fire.

The blessing of the paschal candle, the Exultet, found in the English-language sacramentary contains many cosmic and macrocosmic images of creation. This preconium could be enhanced, however, with the inclusion of some microcosmic images of nature. For instance, the present English translation of the Exultet has eliminated the references (still present in the Latin text) to the working of bees in the creation of wax. While not advocating that the blessing prayer retrieve the image of bees, the question still arises, how in this solemn blessing of light does the evocation of both the macrocosm and the microcosm find a voice in Easter praise?

The blessing prayer for baptismal water in both the Latin- and English-language sacramentaries has been voided of most cosmic and nature symbolism, present in former texts (Stock 1985). In the proclamation of this prayer, it seems that we have a moment to utter the strength of water and its life-giving and death-giving power, as well as the images of creation that are associated with chaos and life. If such symbolism were worked into the blessing, it would more clearly speak of the significance of water and the depth of cosmic indebtedness humans have to its power.

In view of these considerations, the paschal night, indeed, could become the night of new creation, in which Christians celebrate the mystery and wonder of creation and re-creation in Christ: the lighting of the sun, the eruption of water, and the dwelling of light in the midst of darkness.

2. *The Celebration of Christmas.* The Christmas season, too, has potential for a powerful celebration of earth and cosmos. Our attention is first drawn to the celebration of the Eucharist at midnight on December 25. Why *midnight* Mass? Unlike the paschal vigil, which is celebrated as an all-night awakening to the Easter morn, the midnight Mass of Christmas is celebrated at the height of darkness and night. The antiphon, "Dum medium silentium tenerent omnia," from the first vespers of the Sunday following Christmas, could offer a direction celebrating the night and birth of light. The entire antiphon reads in an English translation,

> While all things were poised in the heart of silence, and the night was in the middle of her travel, Your Almighty Word, O Lord, leapt down from the royal heavenly throne, alleluia!

The midnight image and the proclamation of light within it provides another appropriate occasion for the inclusion of cosmic images. Along with this, the Christmas season offers the symbolism of the Star of Bethlehem, sought out by the astrologers. This image could also be worked into the prayers and blessings of this night.

Lastly, the Christmas season, in many places of the world, is commonly celebrated with greenery (e.g., trees and wreaths). What blessing of earth can be associated with this custom, and associated with the mysteries of creation and redemption? In this popular practice, could a development of blessing prayers evoke the cosmic symbolism of tress, which unite earth and sky?

B. The Body and Symbolic Activity

A second liturgical field in which the insights of a renewed cosmological consciousness may be pertinent is the *bios* of ritual, that is the bodily need of the liturgy itself. It is clear that poetry and narrative are connected with ritual and body, as in the relationship between matter and form in traditional sacramental theology. Therefore, in order that the renewed creation story may find root and appropriation in the life of the Christian community, there must be a renewed attunement to the body. This attunement would find expression in dance, gesture, song, and in the construction of spaces for liturgy that would allow and promote such ritual activity. If the narrative does not illumine a bodily reality, a new way of moving in creation, a new way of seeing the moon, a new way of wonderment in the face of light, it will remain empty form. The symbolism of the feasts, embodied in ritual activity, expands the imagination and speaks of the mystery celebrated.

An example may be helpful. At the the paschal vigil the celebration of the paschal fire and candle is more often than not absurd and trivial. The fire is inadequate, the candle is processed into the assembly hall and then the electric lights are turned on. We are rationally symbolic, but even in the best of parishes, has any deep awe and appreciation of light been experienced? The lighting and blessing of the paschal fire and candle emerged from the practical need to have light. It was from this basis that light developed its symbolic meaning. The question becomes then, how can we relate the blessing of light to today's measures of supplying light, and enlarge the cosmic symbolism of these elements of the vigil?

Do the eyes and bodies of the participants pulse with wonderment at the Milky Way, the moon, the sun, and the vast galaxies of light of which we are a part? If electricity is used, does it invite a wonder at the power of technology and the wonders of hydroelectric sources of energy? The point is not to contemplate the data of science, but to use light and gestures with fire and technology in a way that would evoke a wonderment, not of what humans can create and do, but of the awesomeness, energy, and power of light itself, however generated. How this is to be done in an assembly is yet to be discovered. All that can be said, however, is that the service of light is presently banal, even while rubrically correct.

Perhaps some awakened creativity in regards to light may come from secular celebrations. On occasions of great festivity, such as the inauguration of a president or Independence Day, the United States has a celebration that awakens a visceral sense of the stupendous power of light and the awesome dimensions of technology and science. The display of fireworks speaks to the body and the eyes the wonder of light. It is not so much a celebration of what we have made, as a celebration of the manifestation of the power of light, a joyful enchantment in brilliance.

While not necessarily recommending that we use fireworks at the paschal vigil, the question remains, how does the liturgy use the fullness of symbol, of sign and body, to bring the community into a state of wonder at the wild dimensions of the universe and the mystery of light. In the midst of this wonder, evoked in the body, the proclamation of creation is given new psychic energy. In the liturgical assembly it is then related to Christ, the bright morning star who shines and sheds light on all creation.

III. Conclusion

This article began with questions which emerged from an original Eucharistic Prayer text. It is fitting to end at the same juncture, for Eucharistic Prayer A points the way towards an integration of cosmological symbolism and creation narratives into Christian liturgical prayer. Ong, Berry, and Swimme, it was claimed, support and give certain directions for this endeavor. We should not give up on the venture, because its implications are too significant to ignore. This first effort of ICEL heralds the beginning of a vast project, yet to be undertaken.

In the end, this attempt at an original composition points us to the ever-present and central reality of knowing gratitude for creation in the blessing of bread and wine at the Eucharistic table. In these great cosmic symbols of food and drink, earth, sky, God, and mortals are drawn into communion, and the heart of the liturgy reveals itself as the heart of the galaxies, all for the glory of God.

REFERENCES

Berry, Thomas
1988 *The Dream of the Earth.* San Francisco: Sierra Club Books.

International Committee on English in the Liturgy, Inc.
1984 *An Original Eucharistic Prayer: Text 1*
1986 *Eucharistic Prayer A*

Ong Walter J., S.J.
1991 "God's Known Universe and Christian Faith—Pastoral, Homiletic and Devotional Reflections." *Thought,* 66, no. 262 (September) 241–58.

Stock, Alex
1985 "The Blessing of the Font in the Roman Liturgy." *Concilium* 178 (April, 1985), eds. Mary Collins and David Power, 43–52.

Swimme, Brian
1987 "Science: A Partner in Creating the Vision." In *Thomas Berry and the New Cosmology,* eds. Anne Lonergan and Caroline Richards, 81–90. Mystic, Conn.: Twenty-Third Publications.

Swimme, Brian, and Thomas Berry
1992 *The Universe Story: From the Primodial Flaring Forth to the Ecozoic Era—A Celebration of the Unfolding of the Cosmos.* San Francisco: Harper Collins.

7

The Preparatory Rites: A Case Study in Liturgical Ecology

Edward Foley, Capuchin, Kathleen Hughes, R.S.C.J., and Gilbert Ostdiek, O.F.M.

Introduction

This essay joins a series of recent assessments of the preparation of the table and the gifts. Thomas Krosnicki offered an interesting analysis of these rites (Krosnicki 1991). He was concerned to underscore the modest, preliminary nature of these rites, insisting that all misleading notions of offering or "little canon" be removed, leaving the offering as a constitutive element of the anaphora. To that end he suggested a simplification of texts and a rearrangement of the structure. Frederick McManus has also provided a recent assessment of the preparation of the gifts (McManus 1990). He analyzed the evolution of these rites in the work of the Consilium from 1964 to 1969, evaluated the final shape of the rites, and offered suggestions for the future. Cultural adaptation is one of his major concerns: a development that appears more possible if the rite to be adapted has achieved a noble simplicity and, thus, is more susceptible to cultural elaboration.

The questions we bring to a study of the preparatory rites are of a different sort. The last few decades have been a time of growing interest and concern about our environment. The extinction of plant and animal species, the pollution of our water supply, and the depletion of critical resources have generated a new consciousness about our biosphere. With this awareness has come the slow development of social and political policy about our environment. Developing even more slowly is an adequate theology of creation that addresses the current state of the biosphere. Prior to 1980 very little of a serious theological nature was written on this topic. In contrast the 1980s witnessed the publication of numerous theological treatises on the relationship between human beings and the earth (notable contributions include: Winter 1981; Gustafson 1982; Jonas 1984; Moltmann 1985; Granberg-Michaelson 1987; Berry 1988). Even the authors of such works, however, acknowledge that they are only in the initial stages of this inquiry.

While there are many paths that one could follow in exploring our relationship to the created world, the liturgical bias is to discover how the belief enacted in worship communicates this relationship. The *locus classicus* of this notion is the maxim of Prosper of Aquitaine, "legem credendi lex statuat supplicandi."[1] While this phrase can be interpreted in a number of ways (see Irwin 1990, 15 n. 2), the original statement suggests that liturgy is a source of theology to the extent that it is rooted in Scripture and is the expression of a praying Church. This traditional belief was reaffirmed in Vatican II's Constitution on the Sacred Liturgy (1963) which teaches that the Church's liturgy, especially the Eucharist, is the font and summit of the Church's life (no. 10). Thus the Church's public worship is not only a source of personal belief, but the foundation of shared belief. It is from this perspective that the liturgy can be called the Church's first theology. In many respects the maxim of Prosper and the teaching of the Constitution on the Sacred Liturgy are confessional expressions of a transcultural phenomenon, namely, that people's beliefs find their fullest expression in ritual.[2]

Given the centrality of the Eucharist in Roman Catholic practice and thought, it seems appropriate to begin this inquiry by reflecting on the ecological stance embedded in the Church's Eucharistic liturgy. To focus the discussion even further, we will limit our remarks to the preparation of the table and the gifts. Specifically, this essay will explore the elements, gestures, and texts employed during these rites in order to discover their implied theology. To achieve a clearer picture we will compare the elements, gestures, and texts in the Missal of Paul VI (revised *editio typica,* 1975) and those of the Missal of Pius V (1570). This comparative approach will enable us to highlight the directions of current liturgical reform, to note some of the challenges of the reform yet to be addressed, and to suggest a direction for the further reform of this rite.

The Elements

An Historical Note

From the beginning of Christianity the bread used at Eucharist was ordinary table fare. After the ninth century unleavened wheat bread became customary and eventually mandatory in the Christian West. Alcuin (d. 804) offered one of the first testimonies for this transition: "The bread which is consecrated into the body of Christ ought to be most pure without leaven or any other additive."[3] There was an ancient tradition for making Eucharistic bread in the shape of flat disks. Bread stamps of the period indicate that these disks were larger and thicker than the hosts of a later era. The new restrictions against leaven or other additives produced Eucharistic bread that was round and flat.

The reasons for this change were many and complex. One factor was increased emphasis on the unworthiness of the laity: unworthy people did not go to communion, nor did they bring gifts of bread and wine to the Church as an offering. Consequently there was not only a significant decline in the number of communicants but also in offertory processions. Where the offertory procession did survive the people's gifts were not used directly in the Mass. Increasingly the elements employed in the Eucharist were prepared by monks or clerics. Growing emphasis on the otherworldliness of the Eucharist led to the use of bread that was different from ordinary table fare. Understanding Eucharist as the reenactment of the life and death of Jesus lent further support to the use of unleavened wheat bread since medievals commonly believed that Jesus used such at the Last Supper.[4] Thus at the turn of the millennium ordinary table bread gave way to hosts. An increase in the number of private Masses coupled with a decline in the number of communicants contributed to a reduction in the size of the host that often was consumed by a single presider. Another factor was the practice, evident since the ninth century, of placing the bread on people's tongues instead of in their hands. By the eleventh century small hosts were prepared for the few remaining communicants and the use of unleavened bread became universal in the West. So did the practice of baking hosts of two separate sizes—a larger one for the priest and smaller ones for lay communicants. As early as the fourteenth century the baking of hosts had, in some places, become a business that required special ecclesiastical approval.

Early Christians not only employed ordinary table bread for their Eucharists but also ordinary table wine. Hellenized Judaism and the Roman Empire presumed that wine was a product of the grape. This presumption was so clear in Christianity that few documents from the early centuries needed to acknowledge it. One exception is the Fourth Council of Orleans (541) which noted, "One may not presume to offer anything else in the oblation of the holy chalice except that which is the fruit of the vine, and this mixed with water."[5] By the later Middle Ages there was more discussion of the essential matter for the sacrament. Eugene IV reiterated the traditional practice: "The matter of this sacrament is wheat bread and grape wine with a small amount of water,"[6] a view repeated at Trent (1545–63).

Over the centuries various authors and sacred congregations have raised questions about the essential qualities of wine for the Eucharist, including questions about frozen wine, wine made from wild grapes, wine produced from something other than grapes, alcoholic content, and additives to the wine. Few canonical decisions about these issues have emerged except regarding the alcoholic content of the wine. In 1910 the Holy Office did state that "wine from which all the alcohol has been extracted by human means is neither valid nor licit matter for consecration."[7] On the other end of the spectrum, the Holy Office has set the limits of alcoholic content at 12 percent for dry wines and 18 percent for sweet wines (see McSherry 1986, 44). The granting of various indults by the Holy Office from 1960 on for the use of unfermented grape juice in the Eucharist, however, has rendered much previous opinion on alcoholic content moot. The medieval presumption of grape wine continues within the Roman documents.

The Elements in the Missals of Pius V and Paul VI

The medieval tradition was incorporated into the Missal of Pius V, which required that bread be wheat for the sake of validity[8] and unleavened for the sake of liceity.[9] Wine must be "of the vine," that is, grape wine, both for liceity and validity.[10] The Missal of Paul VI offers a mixed message on the nature of the elements to be employed in the Eucharist. The *General Instruction of the Roman Missal* (GIRM) notes, "The nature of the sign demands that the material for the Eucharistic celebration truly have the appearance of food. Accordingly, even though unleavened and baked in the traditional shape, the Eucharistic bread should be made in such a way that in a Mass with a congregation the priest is able actually to break the host [sic] into parts and distribute them to at least some of the faithful" (no. 283). This message is further clouded by the instruction *Inaestimabile Donum* (1980) which requires that bread for the Eucharist be baked only with flour and water. As to the wine, the *General Instruction* notes that "The wine for the Eucharist must be from the fruit of the vine, natural, and pure, that is not mixed with any foreign substance" (no. 284).

An Interpretation

What is most striking about the traditional Eucharistic elements of bread and wine is that neither are found in nature; rather, both require a transformation of nature. Berries and spring water do not provide us with the elements for Eucharist which requires gift of the earth, fruit of the vine, and *work of human hands.* The cooperation between nature and humanity is implicit in the very elements of bread and wine. Furthermore, this cooperative venture presumes a long and reciprocal relationship. Bread is not produced overnight. The land needs to be cleared, seeds planted, crops tended, grain harvested, milled, and baked. Then the land rests before the cycle begins again. And while a grain crop can be produced in a matter of a few months, the reciprocal relationship required of producing wine is even greater. Vines must be planted, tended, and pruned for years before they produce a grape of any quality. Even after the harvesting of the grape, the produce needs to ferment and mature before achieving the dignity of wine. Thus, the very elements of bread and wine presume a symbiotic relationship between humankind and nature, especially for the sustained production of these elements.

When Christians employed table bread and table wine for Eucharist, the ritual implicitly acknowledged the need for people to cooperate with nature and with God. It was this tripartite relationship that enabled grain and grapes to be grown, bread and wine to be produced. One might suggest, however, that the progressive employment of material unrecognizable as bread, coupled with the withdrawal of the cup from the community, contributed to the erosion of this cooperative relationship. Christians used out-of-this-world bread (hosts), prepared by out-of-this-world people (monks and nuns), employed in an out-of-this-world ritual (Eucharist). In this "otherworldly" approach, virtually all ties with nature were severed: Eucharistic bread and wine were removed from the food chain, and almost from the realm of food. This withdrawal is epitomized in the production of hosts whose shelf life far exceeded that of any other "bread": hosts that were not consumed by people as much as they were stored in tabernacles or encased in monstrances. Thus the final withdrawal, in which bread ceased to function as nourishment for the body and, instead, became fascination for the eye.

Removed from the world of human interaction, produce, and nourishment, hosts and a few drops of "sacramental wine" became the stuff of which sacrifices were made. The gifts were obliterated in the ritual texts and in the private consumption of the priest. Anthropologist René Girard contends that conflict is a fundamental element of human existence, and ritual sacrifices serve to divert one's own latent violent tendencies to an exterior object (Girard 1977).[11] They are the ultimate act of scapegoating and foundational for many of the world's religions. Could it be that our neutralization of bread by transforming it into otherworldly hosts gave us the implicit permission to sacrifice not only the host but also the physical world in order to satisfy our own violent tendencies? Artificial food, like hosts, is so far removed from nature that we have become immune to the ritual consumption and destruction of such produce. So have we become immune to the implicit affirmation of our instinct and ability to sacrifice nature for our own needs.

Furthermore, the Western bias for hosts and grape wine as the only acceptable materials for Eucharist symbolizes not only the dominance of a specific culture but also of a specific ecosystem. Lifted out of the geo-system, this staple food and festive drink is upheld as the only staple food and festive drink worthy of divine presence. The primacy of these elements implicitly asserts a primacy about the ecosystem that produced them. It is not a great leap from such a perspective to a rationalization of the rape or destruction of other ecosystems that appear to be both inferior and unconnected to our own.

The Missal of Paul VI made a halting effort to shift from out-of-this-world bread and wine to the work of human hands, though it offers no challenge to the Western bias of what the resulting produce should look like. It is possible to suggest that the shift to "work of human hands" is also an implicit move from scapegoating to symbiosis; a nudge from consumerism to cooperation; a first step away from an instinct to sacrifice nature for the sake of some distant divinity to cooperation with nature wherein the divinity is at least partially revealed. Maybe this is the true hope of shifting from "offertory" to "preparation of the table and the gifts" wherein we no longer proclaim our dominance over nature or our ability to develop non-bread mutations of bread for our sacrifices, but assert our need for cooperation in the triad of divinity, creation, and humanity. We have yet to learn in the process, however, how to value such cooperation in ecosystems other than our own.

The Gestures and Words

To achieve a fuller reading of the ecological issues already raised by the elements themselves, it helps to set them in the context of the gestures and words of the Eucharistic rite. Anthropologically, food-language consists not only of food and drink, but of the entire chain of human actions involved in their production, distribution, preparation, and consumption. Similarly, the theological meaning of Eucharistic food is to be found not only in the elements, but in the whole complex of actions and words surrounding the assembly's sharing of them. The preparation rite is one part of that ritual complex.

An Historical Overview

The Jewish precedent for our "preparation of the gifts" was a very simple action. The foodstuffs were brought to the table before the meal. As the meal began, the head of the household took the bread and held it in preparation for saying a blessing; at the end of a festive meal the same action was repeated with a cup of wine (Matt 26:26-28 and par.). The prayer of blessing expressed the religious meaning of taking and sharing the food and drink. The preparation rite evolved from this simple action of bringing the food to the table and taking it up for the prayer of blessing (for a fuller account, see Dix 1964, 48–102). Through the course of history this action became more complex and its meaning increasingly obscure. Elaboration of the gestures usually came first; words were gradually added to help express the religious significance of the gestures. The mid-second-century rite described by Justin Martyr was still simple in form: bread and a cup of wine mixed with water were brought to the one who presided.[12] The pattern recorded by Hippolytus in the early third century had become somewhat more elaborate. The deacon brought up the bread and a cup of wine mixed with water, and the presiding bishop, together with the entire presbytery, laid hands on this "oblation."[13] The *Apostolic Tradition* (no. 20) also notes that all the baptized were expected to bring elements for the oblation, a practice reminiscent of 1 Corinthians 11 and increasingly witnessed in writings of the patristic period.[14] Initially the gifts brought forward at the preparation rite included only elements for the Eucharist. According to Justin, the collection of gifts for the poor and needy was taken up after communion, rather than as part of the preparation rite.[15]

The next phase in the elaboration of the preparation rite occurred when the community dramatically increased in size after the Edict of Milan (313) and house churches gave way to regular assembly in public basilicas. Larger numbers were to be fed at the Eucharistic table, requiring greater amounts of bread and wine; the Church's rituals were refashioned to fit imperial patterns and new worship spaces. Thus, the simple act of provisioning the community table with the elements now grew into a ritual presentation of these materials in a solemn procession. In some rites the people brought up the gifts; in other rites deacons selected bread and wine from the supply that the people had brought to a side table or a "sacristy" and presented these to the presider. Liturgical ministers and presider added their own gifts. Bringing up and arranging gifts on the altar in Rome during the late patristic period required an intricate choreography, while words remained at a premium. Although at first there seems to have been no prayer to accompany or conclude the ritual presentation of the gifts other than the Eucharistic Prayer—leaving the action to speak for itself—a concluding prayer said aloud over the gifts (*oratio super oblata*) is found in Rome by the mid-fifth century. In that same period a psalm was sung during the procession with the gifts to add a note of festivity and gladness.

In the early medieval period the presentation of gifts became a veritable jungle of ritual actions and prayers. The change from home-style bread to unleavened hosts and the parallel decline in communion by the assembly meant that little bread was needed and that it could not be baked at home. With the eventual withdrawal of the cup, no wine was needed for the assembly. As a result, the people's procession with the gifts eventually disappeared. The bread was now baked by church folk, and it was brought to the altar from the sacristy by the priest or from a credence table by a deacon. The intricate gestures surrounding the physical presentation and offering of the bread and wine were retained, however, making of the rite a clerical action devoid of any active participation by the people, although initially dramatized for their sake. One such dramatization was to reduce the prayer over the gifts to a silent prayer, now called the *secreta.* According to Jungmann, this practice first appeared in mid-eighth-century Frankish territories; it became common in Western liturgy within a century (Jungmann 1959, 2:90–91). In addition, during this period almost every gesture of the rite came to be accompanied by a matching prayer of personal devotion and apology. The *secreta* was only the last of a series of silent prayers by the priest, a silence broken only by the concluding "Per Dominum nostrum."

Two other typical accretions were the *lavabo* and incensation. The washing of hands, rooted in Jewish ablution rites, did not become a constant element of the Christian Eucharist until the early medieval period. Although the washing may have served the practical purpose of cleansing hands after receiving gifts, it had a symbolic meaning from the beginning, since it often occurred before the gifts were received. Jungmann counts the *lavabo* as one of a series of washings, starting with the washing before vesting and continuing at times through multiple washings at the beginning of the liturgy of the Eucharist. These washings served to mark a progressive movement from the profane into the holy of holies where the sacrifice would take place (Jungmann 1959, 2:76–78). Similarly, the gesture of incensing gifts, altar, and people, which dates to the mid-ninth century, served that same ritual purpose of purification.

The act of taking up bread and wine for the table, which formed the original core of the preparation of the gifts, had thus become something very different by the Middle Ages. The actual taking of bread and cup had, in fact, been broken into three moments: the lifting up of the elements from the altar at the "offertory," at the "consecration," and at the communion. The first of these soon evolved into a solemn ritual presentation of bread and wine by people and ministers and then into a highly ritualized, clerical preparation and offering of the gifts performed solely within the sanctuary precincts. The effect of the medieval "offertory" rite was to distance the gifts and the priest from the people and things of the everyday world by creating a series of ritual boundaries.

The medieval preparation rite, consolidated in the Missal of Pius V, was still largely intact at the time of Vatican II. It was the desire of the Consilium to restore the simplicity, clarity, and brevity of the ancient preparation rite by (1) removing useless interpretative words, (2) eliminating proleptic suggestions of offering or sacrifice, typical of medieval accretions, and (3) restoring the role of the assembly so characteristic of the early Church (see Bugnini 1990, 337–92). Their success can be gauged by comparing the Missals of Pius V and Paul VI.

The Preparation Rite in the Missals of Pius V and Paul VI

A comparison of the "Offertorium" (rite of 1570) and the "Preparation of the Altar and the Gifts" (rite of 1975) shows that the Vatican II reform kept the basic pattern of the medieval "offertory," with some significant revisions and reductions.

The procession with gifts. The 1570 rite was performed by the priest with his back to the assembly, except for an opening greeting and a closing invitation. As the "offertory" began, the priest kissed the altar, turned to greet the people, and turned back to the altar to say "Oremus" and the "offertory" antiphon. This antiphon was all that remained of the complete psalm chant which once had been sung during the procession with the gifts. The antiphon, usually a psalm *incipit* or verse, was variable and proper to the Mass of each day. In some Masses a schola or choir sang the "offertory" antiphon[16] which the priest had recited silently. Meanwhile, the priest prepared the altar by unveiling the chalice and paten and unfolding the corporal. He had already brought chalice, paten, and his host from the sacristy to the altar at the beginning of Mass. In solemn high Masses the subdeacon brought the chalice, paten, and the priest's host from the credence table, where they had been placed before the celebration began, and the deacon completed the preparations. In neither case was there a true procession with visible bread and wine.

The 1975 rite begins with servers or a deacon preparing the altar. The rubrics recommend that members of the assembly form a procession to bring up bread and wine in sufficient quantity for communion of that assembly. Other gifts for the Church and the poor may be brought up as well. The gifts are handed to the presider, who is assisted by the deacon or ministers in receiving them. The bread and wine are placed on the altar. An "offertory song" may be begun when the ministers prepare the altar. The function of the song is to accompany and celebrate the communal aspect of the procession of gifts; its contents are determined by feast or season and need not speak of bread, wine, or offering.[17] If a presentation song is not sung, the antiphon is omitted (GIRM nos. 26, 50). Because of the optional and completely variable nature of the sung processional text, little may be inferred about its role in relationship to the gifts other than lending festivity and joy to the procession.

The presentation of the gifts. In the 1570 rite, after taking the paten with the host (from the altar or the deacon), the priest raised it high in an offering gesture.[18] Making a sign of the cross with the host, he placed it on the corporal. Next, the priest blessed the water and he (or the deacon and subdeacon) poured wine and water into the chalice. The priest then raised the chalice in an offering gesture. Making a sign of the cross with the chalice, he placed it on the corporal. Silent prayers accompanied each of these actions.

The presentation prayers were a curious amalgam. The prayer for the oblation of the bread was in the first person singular: the priest offered the "spotless host" in his own name ("which, I your unworthy servant, offer to you") and, in the first place, for his own benefit ("for my own countless sins, offenses, and negligences")[19] as well as for those present and all the faithful, living and dead. The prayer for the offering of the chalice, on the other hand, used first person plural language exclusively and petitioned "for our salvation and for that of the whole world." The laws of comparative liturgy suggest that the latter text is more ancient and that the former was influenced by the same apologetic unworthiness that led to the medieval creation of a whole genre of such texts. Both prayers beseeched God to accept the Eucharistic offering, "this spotless host" and "the chalice of salvation." In the judgment of the Consilium, neither text was an "accurate expression of the genuine meaning of the 'offertory' rites but merely anticipated the meaning of the true and literal sacrificial offering that is present in the Eucharistic Prayer after the consecration, when Christ becomes present on the altar as victim" (Consilium 1970, 37–38). The prayer which accompanied the ancient custom of diluting the wine with water—even though the wine no longer needed cutting like the thick Middle Eastern wine of antiquity—made the custom an allegory of the union of Christ and Christians. The text, adapted from an ancient prayer assigned for Christmas Day, extolled the marvelous exchange between divinity and humanity in the incarnation (Bruylants 1952, 1:13). The presentation of the host and chalice concluded with a series of gestures and prayers humbly asking God to accept us and our gifts, and invoking the Spirit to sanctify the gifts. The priest bowed before the altar, with hands joined on the altar, and prayed, "In humility, with contrite heart, may we be acceptable to you . . ." He then stood erect, extended, raised, and joined his hands in the *circulus* gesture, while raising his eyes to heaven and then lowering them. Then he made the sign of the cross over the gifts and prayed, "Come, O Sanctifier . . ." This entire sequence of prayers and gestures anticipated one section of the Eucharistic Prayer: the offering made in the anamnesis, the prayer of humble acceptance, and the epiclesis.

The 1975 rite is much simpler. The presider takes the paten with the bread, raises it a little from the altar, and inaudibly prays a berakah:

> Blessed are you, Lord, God of all creation.
> Through your goodness we have this bread to offer,
> which earth has given and human hands have made.
> It will become for us the bread of life.

He then places the paten with the bread on the corporal, without any further gesture. The presider (deacon) pours wine and water into the chalice, praying silently, as in the 1570 rite: "By the mystery of this water and wine . . ." The priest takes the chalice, raises it a little from the altar, and inaudibly prays a second berakah:

> Blessed are you, Lord, God of all creation.
> Through your goodness we have this wine to offer,
> fruit of the vine and work of human hands.
> It will become our spiritual drink.

These texts praise God for the gift of nature, itself a blessing as source of life, energy, and strength. The bread and wine are gifts of our Creator and fruit of human industry in harmony and collaboration with nature. The texts suggest the intimate union of cosmic energy and human industry in these simple gifts which will become spirit and life for the community. The presentation of bread and cup continues to be interrupted by the pouring of water into the wine, with an unfortunate shift to the language of mystery—"by the mystery of this water and wine"—in the midst of profoundly earthy realities. According to the revisers:

> The new formularies for the gifts bring out the giving of glory to God, who is the source of all things and of all the gifts given to humanity. They state explicitly the meaning of the rite being carried out; they associate the value of human work, which embraces all human concerns, with the mystery of Christ. The offertory rite, then, has been restored through that explicit teaching and shines forth with new light (Consilium 1970).

Not exactly and certainly not all of the time! The use of texts reminiscent of Jewish table prayers suggests a renewed emphasis on the Eucharist as a meal. The content of these texts is a highly condensed ecological statement of the tripartite relationship of nature, humanity, and God in collaborative alliance. But the rubrics indicate that these texts are said inaudibly; if there is no song, they may be said in an audible voice. The effect of such rubrics is to suggest that the preparation and presentation of bread and wine are not an act of the community; nor would they appear to concern the whole assembly but only the priest. That inference finds support in the *General Instruction* norm which distinguishes presidential texts that are to be spoken in a loud and clear voice from other texts: "the priest does not only pray in the name of the whole community as its president; he also prays at times in his own name that he may exercise his ministry with attention and devotion. Such prayers are said inaudibly" (GIRM 13). After the presentation of the cup, the presider bows and prays quietly, as in the 1570 rite. The older text, "In spiritu humilitatis," has been replaced by an abbreviated but substantially similar statement of humility and contrition. The text also speaks of "the sacrifice we offer you,'' an unfortunate anticipatory reference to the gifts as sacrifice, the kind of language that made the ''offertory'' a ''little canon.''

The incensation. In solemn Masses of the 1570 rite, the priest blessed incense, placed it in the thurible, and incensed the gifts, the altar, and the crucifix. He returned the thurible to the deacon (server), who incensed priest and people. Silent prayers accompanied and interpreted each of these actions. The priest blessed the incense through the intercession of Michael the Archangel and prayed that a sacred exchange might take place: the mercy of the Lord upon the community as the incense ascended. A portion of Psalm 141 that concluded the incensation shifted the voice to first person singular. The effect of this shift from plural to singular voice was to remove the community from the purview of the prayer, or to remove the presider and the gifts to a sacral place, a holy of holies.

The 1975 rite gives the priest the option of incensing the gifts and the altar, but without any prayers to interpret the action. A deacon or minister incenses the people. The incensation has been and continues to function as a sacral designation, a mark of honor to holy things reserved for God.

The washing of hands. In the 1570 rite the priest then washed his hands at the side of the altar while silently reciting five verses from Psalm 26, beginning with ''I will wash my hands among the innocent . . .'' He returned to the middle of the altar, where he bowed with joined hands on the altar and again prayed humbly and in silence that God would accept our offering: ''Receive, O holy Trinity, this offering which we make . . .''

In the 1975 rite the presider washes his hands at the side of the altar, praying quitely. Psalm 26 has been replaced in the reform by one verse from Psalm 51, ''Lord, wash away my iniquity; cleanse me from my sins.'' The meaning assigned to this rite is ''an expression of the [priest's] desire to be cleansed within'' (GIRM 52). When the Consilium debated the value and place of the washing of hands as well as whether it should be done in silence or accompanied by an interpretative word, most members wished to retain this rite because of its symbolism of interior cleansing, although the lay consultors objected: ''We do not see the reason for the *lavabo* at the offertory. The symbolical meaning—purification—has already been satisfactorily expressed by the penitential act'' (Bugnini 1990, 368).

The final prayers. To conclude the 1570 rite, the priest kissed the altar, turned to the people, extended his hands and, in a slightly elevated voice, invited the response with a text dating back to the eighth century: ''Brethren, pray that my sacrifice and yours may be acceptable to God the Father almighty.'' The ministers, or the people, answered: ''May the Lord receive the sacrifice from your hands to the praise and glory of his name, for our welfare and that of all his holy Church.'' The priest, who had turned back to face the altar, added a silent "Amen" and prayed the *secreta* silently, saying the conclusion in a clear voice.

The 1975 preparation rite concludes with the presider standing at the center of the altar, facing the people and, with extended hands, inviting them to pray. The text for this prayer dialogue remains that of the 1570 rite.[20] The rubrics do not specify that the presider answer "Amen" to the people's prayer. The presider then says aloud the prayer over the gifts and all answer "Amen." The function of the *super oblata* is to complete the preparation rite and act as a transition and introduction to the Eucharistic Prayer. The content, according to Emminghaus, is supposed to sum up the theme of the entire preparation, namely, that the gifts express the community's will to give itself, but this will needs to be stirred up anew so that the Eucharist may also be the Church's sacrifice in spirit and truth (Emminghaus 1978, 167). Such a tidy assessment is not always borne out in an examination of the prayers themselves. Too often they seem to suggest that the effect of the Eucharistic Prayer somehow flows back on the offerings, creating something of a pre-consecration. A particularly egregious example (*Missale Romanum,* 333) reads:

> Lord,
> may these gifts cleanse us from sin
> and make our hearts live
> with your gift of grace.

Such inaccuracy and theological ambiguity is found in numerous *super oblata,* a situation often not remedied in the revised *editio typica* of the Roman Missal (1975). At the same time, some of the texts in ICEL's recent revision work do measure up to Emminghaus' description, for example text 219:

> Accept, Lord, our offerings,
> chosen from among your many gifts,
> and let this expression of our reverence
> become the promise of your everlasting redemption.

The theological ambiguity of many of the *super oblata,* the proleptic character of their content, the sometimes dubious nature of the intercessory element, the circumscribed nature of the function of the prayer, and the limited vocabulary from which they draw euchological images all contribute to a less than felicitous conclusion to the rite of preparation of table and gifts. This is even more problematic since the prayer over the gifts with its invitation may well be the only audible prayer in the entire rite.

An Interpretation

From this comparison of the gestures and prayers in the two preparation rites it is evident that the Missal of Paul VI has succeeded in greatly simplifying the rite. But it is also clear that the revision has kept the basic shape of the medieval rite intact and has not excised all of its problematic aspects, some of which have a direct bearing on a liturgical theology responsive to issues of ecology.

First, the preparation rite symbolically enacts a series of ever tighter boundaries around the gifts of bread and wine and who may offer them. We have previously noted that the bread and wine have already been isolated from the ordinary food chain by recipe and by customs and rules of production and handling. The sequence of gestures and words in the preparation rite carries that process further. The elements are brought forward by members of the community and transferred from their hands to the hands of the ministers who, in turn, place them on the altar. The presider takes the gifts in his hands, raising them up from the altar, and says a prayer that, when audible, speaks of offering bread which human hands have made and wine which is the work of human hands. Next the presider washes his hands, an action whose effect may well be to suggest to those with no knowledge of Jewish ablution rites that either the gifts or the givers were not entirely clean, or even that the presider is washing his hands of the gifts. If used, the incensation reinforces the symbolism of purifying the gifts and setting them apart from ordinary use. The final ritual dialogue seals the break with the role of the community as co-offerers; they pray that the Lord accept the sacrifice "at your hands."[21] With that, a barrier of gestures and words has arisen. The gifts have been separated from the ordinary store of food and from the hands that brought them forward, and they are entrusted to other, holier hands for transformation and distribution.

Second, the actions of the rite as a whole serve as a series of formal, dedicatory gestures that enact a transfer of the gifts to a holy realm where they will be transformed. The accompanying prayers are filled with language of sacrifice, offering, and saving effects. Coupled with the step by step isolation of the bread and wine from their earthy and human origins, this might be taken to imply a radical discontinuity between God's creative and saving deeds, between God's action and responsible human action. It might be taken to imply that holiness and salvation are to be found only in another realm. Blessing theology and salvation theology would then be sundered.

Third, the words and gestures of the preparation rite also encode the relationships between the assembly and its leaders. Although the reform has done much to restore the active participation of the assembly, the preparation rite still labors under more than a few vestiges of the medieval hierarchical ordering that left the people passive watchers. They have handed over the gifts and the presider now presents them for transformation and the offering.

If these interpretations have merit, the way in which the preparation rite has been and continues to be enacted cannot be without deeper theological significance. To the extent that it focuses more on the transformation of bread and wine than on their connection to the earth and to human action in the world, the rite invites greater attention to God's presence in the Eucharistic elements than in creation and human history. To the extent that it takes the cycle of gift-exchange that lies at the heart of both meal and sacrifice out of the assembly's hands at the moment when the Church prepares to offer the transformed gifts of bread and wine, the preparation rite may seem to absolve the assembly from any further acts of giving and suggest that justice remains an optional response.

The collection can serve as a concrete illustration. When taken up after the communion, as in the early Church, it has far greater intimations of mission and continued responsibility for the world as a habitat for all the hungry than does a collection taken up as part of the presentation of gifts. In the former practice, those who have received nourishment and life at the table are asked to continue the giving. In the latter, the chain of gift-giving might seem to stop when the gifts are handed over to the presider to offer them up to God. Table sharing is only a moment in, and not the end of, the cycle of gift-exchange. The Eucharist of the future may need to give a much clearer witness that the life bestowed by the earth and human work and nurtured in the meal must be given for others and become productive in turn.

Assessment and Directions for Further Reform

To a limited extent, the goals for the revision of the *offertorium* have been met in the Missal of 1975. Useless repetitions have been removed; many of the formularies have been reduced in length; some anticipation of offering and sacrifice has been eliminated; some of the proleptic language of sacrifice imbedded in medieval accretions has disappeared; and the role of the community has been heightened, particularly in the restoration of the procession of gifts. The revisers appear to have been satisfied with their work:

> History teaches that the offertory rite is an action of preparation for the sacrifice in which the priest and ministers accept the gifts offered by the people. These are the elements for the celebration (the bread and wine) and other gifts intended for the Church and the poor. This preparatory meaning has always been regarded as the identifying note of the offertory, even though the formularies did not adequately bring it out and were couched in sacrificial language. The new rite puts this specifying note in a clearer light by means both of the active part taken by the faithful in the presenta-

> tion of the gifts and the formularies the celebrant says in placing the elements for the Eucharistic celebration on the altar (Consilium 1970).

Others are less sanguine about the success of this reform. Some would say that the entire preparation rite still reflects the explosion of ritual gesture and texts that took place from the ninth to the sixteenth centuries. The earliest additions to the unadorned bringing of the gifts, namely, the procession accompanied by a chant and concluded by a simple prayer, produced a structure that remains obscured by a multiplicity of prayers and actions. The community procession and presentation of gifts is optional. The ritual action of taking and raising the gifts slightly as a mark of preparation is more often than not transmuted into a large gesture of offering. Those texts which best capture an ecological spirituality are lost if inaudible. Furthermore, the use of an inaudible recitation excludes the community from this ritual action since ''inaudible'' is a code word in the revised rubrics for texts of personal devotion. The prayers of apology and personal unworthiness have not been suppressed despite the fact that their tone is dramatically dissonant from that of the berakoth in praise of God's goodness and our role as co-creators.

As pointed out earlier, a community's belief system is enacted in ritual. The obverse is equally true: a community's ritual shapes its beliefs. In the case of the rite of preparation, a very schizophrenic ritual now shapes the way the community thinks about creation, its role in the preservation or sacrifice of creation, nature as gift and task, and creaturehood as co-creating with the divine.

''Structural reconstructionism will undoubtedly be on the agenda for the future.'' This is a typically understated ICEL summary of the present state of the inadequately reformed rite of preparation. If ecological spirituality plays a significant role in the shaping of this rite in the years ahead, the following should be considered:

1. The choice of elements integral to the local ecosystem and truly nourishing as food;
2. Community participation in preparing and providing these elements;
3. Restoration of the procession of gifts at all celebrations;
4. Reception of the elements by the ministers in a way that does not remove the gifts or the givers from the ensuing words and actions;
5. Revision of the berakoth translations so that they maintain and highlight the collaboration of nature, humanity, and divinity in these symbolic gifts while reworking the language of offering;
6. Normative recitation aloud of the berakoth;
7. The symbolic placing of the gifts on the altar, without lifting them in any pre-consecratory gesture of offering;

8. Elimination of all "inaudible" prayers and gestures of private devotion such as the *lavabo* that create boundaries and separate the community from this ritual, and reduce this rite to a private ministerial action;
9. A dramatic pruning of the number of *super oblata* texts and the recasting of the invitational dialogue of this prayer so that the *super oblata* functions as a simple conclusion to a profoundly simple action: the bringing of this-worldly-food, by a this-worldly-community, for our good and the good of all this world.

NOTES

[1] *Praeteritorum episcoporum sedis Apostolicae auctoritates de Gratia Dei* 8 (PL 51:209). For a further discussion of this concept, see DeClerck, 1978.

[2] "For it is in ritual . . . that this conviction that religious conceptions are veridical and that religious directives are sound is somehow generated. . . . Whatever role divine intervention may or may not play in the creation of faith . . . it is, primarily at least, out of the context of concrete acts of religious observance that religious conviction emerges on the human plane" (Geertz, 1973:112–13).

[3] Alcuin, Letter 90 to the brothers in Lyons (798 A.D.) (PL 100:289).

[4] While the tradition that Jesus employed unleavened bread at the Last Supper is probably accurate, the historicity of the wheat tradition is less tenable. It is possible that Jesus and his disciples used the flat barley cakes of the Passover festival in their last meal together (see Foley, 1991:3–24).

[5] Can. 4 = Mansi 9:114.

[6] Letter of Eugene IV to the Armenians, 1438 A.D. (Mansi 31:1056).

[7] The text of the response appears neither in the *Acta Apostolicae Sedis* nor in *Codicis Iuris Canonici Fontes*. The decision is cited in Noldin-Schmitt, 1925:III:112, n. 4.

[8] "Requiritur enim, ut sit panis triticeus," *De defectibus materiae* 2.

[9] "Si non sit azymus, secundum morem Ecclesiae Latinae, conficitur, sed conficiens graviter peccat," *De defectu panis* 3.3.

[10] "Requiritur enim ut . . . vinum de vite," *De defectibus materiae* 2.

[11] For divergent interpretations of Girard's theory of sacrifice vis-à-vis liturgical theology, see Agnew, 1987, and Gittins, 1991.

[12] Justin, *Apology* 1.65.

[13] Hippolytus, *Apostolic Tradition,* 4 and 21.

[14] See, for example, Cyprian, *De Opere et Eleemosynis,* 15 (CSEL 3A:64–65).

[15] *Apology* 1.67.

[16] Before Vatican II, under the influence of German liturgical renewal, an "offertory song" was often sung instead. The text of this song usually focused on the Eucharistic offering.

[17] *Music in Catholic Worship,* no. 71; ironically, while providing that this song may accompany the procession and preparation of the gifts, the document lists it under "supplementary" rather than "processional" songs.

[18] This was one of the few gestures that was not shielded by the body of the priest and could be seen by the congregation with some clarity.

[19] English translations of the Latin prayers from the 1570 rite are taken from *The Maryknoll Missal* (New York: P.J. Kennedy & Sons, 1966).

[20] The ICEL revision underway is considering a simplified invitation and response based on the model found in the French Sacramentary.

[21] This image resonates with the pseudo-epiclesis in Eucharistic Prayer 1, that the offerings may be carried by the hands of the holy angel to the altar on high.

REFERENCES

Agnew, Mary Barbara
1987 "A Transformation of Sacrifice: An Application of René Girard's Theory of Culture and Religion." *Worship* 61:493–509.

Berry, Thomas, ed.
1988 *The Dream of the Earth.* San Francisco: Sierra Club Books.

Bruylants, Placide
1952 *Les Oraisons du Missel Romain: Texte et Histoire.* Etudes liturgiques 1. 2 vols. Louvain: Abbaye du Mont César.

Bugnini, Annibale
1990 *The Reform of the Liturgy, 1948–1975.* Trans. Matthew J. O'Connell. Collegeville: The Liturgical Press.

Consilium
1970 "Documentorum Explanatio." *Notitiae* 6:37–38.

DeClerck, Paul
1978 " 'Lex orandi, lex credendi,' Sens originel et avatars historiques d'un adage équivoque." *Questions liturgiques* 59:193–212.

Dix, Gregory
1964 *The Shape of the Liturgy.* London: Dacre Press.

Emminghaus, Johannes H.
1978 *The Eucharist: Essence, Form, Celebration.* Trans. Matthew J. O'Connell. Collegeville: The Liturgical Press.

Foley, Edward
1991 *From Age to Age.* Chicago: Liturgy Training Publications.

Geertz, Clifford
1973 *The Interpretation of Culture.* New York: Basic Books.

Girard, René
1977 *Violence and the Sacred.* Trans. Patrick Gregory. Baltimore: Johns Hopkins University Press.

Gittins, Anthony
1991 "Sacrifice, Violence, and the Eucharist." *Worship* 65:420–35.

Granberg-Michaelson, Wesley, ed.
1987 *Tending the Garden.* Grand Rapids: Eerdmans.

Gustafson, James
1982 *Ethics from a Theocentric Perspective.* Chicago: University of Chicago Press.

Irwin, Kevin
1990 *Liturgical Theology: A Primer, American Essays in Liturgy.* Collegeville: The Liturgical Press.

Jonas, Hans
1984 *The Imperative of Responsibility: In Search of an Ethics for the Technological Age.* Chicago: University of Chicago Press.

Jungmann, Josef
1959 *The Mass of the Roman Rite: Its Origins and Development.* Trans. Francis Brunner. 2 vols. New York: Benziger.

Krosnicki, Thomas A.
1991 "Preparing the Gifts: Clarifying the Rite." *Worship* 65:149–59.

McManus, Frederick R.
1990 "The Roman Order of Mass from 1964–1969: The Preparation of the Gifts." In *Shaping English Liturgy,* eds. Peter Finn and James Schellman, 107–38. Washington: Pastoral Press.

McSherry, Patrick
1986 *Wine as Sacramental Matter and the Use of Mustum.* Washington: National Clergy Council on Alcoholism.

Moltmann, Jurgen
1985 *God in Creation: A New Theology of Creation and the Spirit of God.* San Francisco: Harper & Row.

Noldin, H., and A. Schmitt
1925 *Summa Theologiae Moralis.* 17th ed. Oeniponte: Rauch.

Winter, Gibson
1981 *Liberating Creation: Foundations of Religious Social Ethics.* New York: Crossroad.

Part IV

Perspectives from Spirituality

8

Ecology and Spirituality: A Twelfth-Century Perspective

Thomas A. McGonigle, O.P.

Although the ecological challenge has assumed new significance in the last decades of the twentieth century, the relationship between the economic need to develop natural resources and the social responsibility to protect the environment is a perennial question. At moments of great crisis in our personal and communal lives, we look to a variety of sources, including history, to find some experiential wisdom that we hope can provide new insights to help us in confronting the problems we face. This essay seeks to look at the question of the relationship between the natural world and human development from the perspective of one of the great theologians and spiritual writers of the twelfth century, Hugh of St. Victor (1096–1141).

The Twelfth Century Vision of Nature and History

The cultural and intellectual renaissance of the twelfth century marked a new awakening to the reality of the natural world. For the first time in western Europe since late Antiquity, there was a new interest in an active discovery of the workings of nature and the impact nature had upon human life (Ladner 1982, 10). The new naturalism of the sculpture program at the great Romanesque cathedrals of Autun and Chartres and the writings of the masters of the school of Chartres (Bernard [d. 1130] Thierry [d. after 1151], and Gilbert de la Porree [1080–1153]), reflect this new sensibility to nature.

This new awareness of nature involved a deep sense of the connectedness of all created reality: "The order and disposition of all things from the highest even to the lowest in the structure of this universe so follows in sequence . . . that of all things that exist none is found unconnected or separable'' (Hugh 1951, 29). Interest in nature as a phenomenon that could be investigated by reason did not diminish the respect of twelfth-century thinkers for the sacramental character of the universe as sacred and revelatory of God. Reason, guided by faith, could in fact enable one to discern more clearly the beauty and perfection of the *vestigia Dei* (the traces of God) in every aspect of the natural world.

The human person as the *imago Dei* (the image of God) was seen as the link between the natural world and the spiritual world. Human persons would be assisted in knowing the truth of human destiny within the divine plan by reflecting on the natural world from the perspective of faith. But the natural world needed to be understood not as a static reality, but as the place where the destinies of human beings unfolded in the great drama of salvation history, involving the interplay of both divine and human freedom. For the theologians and spiritual writers of the twelfth century, nature and history were tied together in a great plan of salvation that extended from creation to the eschaton.

The goodness of the natural world, coming from the creative act of God, had, indeed, been wounded by the alienation that followed from human sin, but it also shared in the new creation that flowed from the redemptive activity of the incarnate Word. Redeemed men and women were meant to be partners with God in humanizing the natural world through the arts and sciences, while respecting and living in harmony with nature (Chenu 1968, 45).

The great themes of nature and history, creation and redemption, sin and grace, the sacraments and spirituality which occupied such a prominent place in the theological renaissance of the twelfth century, found an all-embracing and coherent presentation in the writings of Hugh of St. Victor, an Augustinian canon, who directed the school at the abbey of St. Victor in Paris from 1133 until his death in 1141. In the exposition of Hugh's thought that follows, we will see how a twelfth-century theologian tries to make sense of the human journey to the fullness of life with God in a world newly aware of nature and history. In Hugh's time, technological advances were being made, and men and women yearned to find the reality of God through a vision of Christian life that included both contemplation and active involvement in the world around them.

Foundation and Restoration

For Hugh of St. Victor the divine activity in relation to created reality consists of the work of foundation and the work of restoration: ''Therefore, the work of foundation is the creation of the world with all its elements. The work of restoration is the Incarnation of the Word with all its sacraments, both those which have gone before him from the beginning of time, and those which come after, even to the end of the world" (Hugh 1951, 3). Adam and Eve were created to reflect the infinite perfection of the divine Trinity within themselves. Yet this divine work of foundation was disrupted because the first human beings disobeyed God's plan of salvation and chose to commit sin.

Hugh likens the condition of fallen humanity to the state of darkness. The opening chapter of Genesis, wherein creation moves from the state of darkness to the state of light, constitutes for Hugh a "sacrament" of the work of restoration. Fallen humankind must pass from the darkness of sin to the light of wisdom. Like the chaos and darkness of the beginning, sinful human beings would remain unformed unless the divine Wisdom provided order and light through a new divine initiative, the work of restoration. The incarnation of the Word provides the remedy for restoring men and women to the vision of God and God's creation.

The work of restoration which requires the renewal of the *imago Dei* at the innermost core of the human person takes place within history as the *ordo rerum gestarum,* the succession of deeds done in time. Since Hugh emphasizes that the restoration of humanity occurred within an historical context, it is understandable that he is concerned about the economy of salvation and the development of universal destiny. The work of restoration is the journey of humanity, which returns to God through the activities of the divine Word-Wisdom.

Symbols/Sacraments

Hugh understands the visible *mirabilia Dei* as symbols of the invisible *mirabilia Dei.* The great Victorine sees all things in history as sacramental. Natural objects and historical events convey spiritual meaning by signifying divine realities.

The human person stands between the visible and invisible worlds: "The first and principal mystery (*sacramentum*) of the divine Wisdom, therefore, is the created wisdom, that is, the rational creature, which partly visible, partly invisible, has been made its own gateway and road to contemplation" (Hugh 1961, 177). Within the Hugonian system of thought, symbols are the means by which the created wisdom, the rational creature, through contemplation finds its way to its ultimate source, the invisible God. Hugh offers the following definition: "A symbol is a juxtaposition that is a cooptation of visible forms brought forth to demonstrate some invisible matter" (cited in Chenu 1968, 103).

The symbolic dimension of human experience was of tremendous significance for the men and women of the twelfth century. Living as we do in a technologically advanced civilization, where nature is the object of scientific analysis and control, and where historical events are seen as the concurrence of social, political, economic, and religious factors, it is difficult for us to share in the experience which our medieval forebears knew in the presence of nature and history. Yet, if we are to understand Hugh's vision of Christian life, we must try to enter into the medieval outlook: ". . . the conviction that all natural or historical reality possessed a *signification* which transcended its crude reality and which a certain symbolic dimension of that reality would reveal to the human mind. . . . Giving an account of things involved more than explaining them by reference to their internal causes; it involved discovering that dimension of mystery" (ibid., 102).

The symbolism of nature and history merged in the liturgy. Here water, wine, bread, oil, and salt, everyday things of the natural world, became the means for entering into communion with the divine. The events of sacred history recorded in the Scriptures also became present reality through their proclamation and celebration in the divine office and the Mass. This ritual representation of historic mystery could draw upon the sacramental character of nature, since even before it was consciously incorporated into any symbol system, the sacramental universe was filled with God.

Since human beings need symbols in order to return to the full light of wisdom, throughout history symbols/sacraments are adopted and refined by the divine initiation to provide a progressive revelation of truth. The ascent of individual men and women to restored divine life occurs within the matrix of the human race's restoration to its divinely willed destiny. Hugh describes the first steps of the historical restoration of humanity as follows:

> A great chaos of oblivion overwhelmed the human mind; humans were sunk in profound ignorance and did not remember their origin. Yet a spark of the eternal fire lived in them, and by its light, as of a spark in the darkness . . . they could seek what they had lost . . . and they began to escape the evils from which they suffered by the study of wisdom. Hence there arose the theoretical sciences to illuminate ignorance, ethics to strengthen virtue, and the mechanical arts to temper human infirmity. (Cited in Southern 1971, 170)

Rooted deeply in the Augustinian tradition, Hugh believed "that the human mind is made in the image of divine Wisdom, and that this image, damaged by the fall, is to be restored through the arts, in cooperation with grace" (Hugh 1961, 24).

Restoration within History

In the Hugonian vision, historical restoration parallels historical creation: "Therefore, the works of foundation, as if of little importance, were accomplished in six days, but the works of restoration cannot be completed except in six ages. Yet six are placed over against six that the Restorer may be proven to be the same as the Creator'' (Hugh 1951, 4). Hugh took this historical framework of the six ages of salvation history, formulated by Augustine and refined by Venerable Bede, as the basis for his own systematic presentation of Christian doctrine, since it enfleshed the three periods of human history: the natural law, the written law, and grace.

The six ages are as follows: (1) from Adam to the flood; (2) from the flood to Abraham; (3) from Abraham to David; (4) from David to the Babylonian captivity; (5) from the Babylonian captivity to Christ; (6) from Christ to the end of time (Hugh 1951, 295–97). Hugh emphasizes that ''from the beginning never have the sacraments of salvation which preceded for the preparation and for the sign of redemption which was completed in the death of Christ been lacking'' (ibid., 296). He summarizes his understanding of the concept of the restoration of humanity within history as follows: ''Therefore, three things occur here for consideration in regard to human restoration: time, place, remedy. The time is the present life from the beginning of the world even to the end of the world. The place is this world. The remedy consists in three things: in faith, in the sacraments, in good works'' (ibid., 142).

Christ the Light-Wisdom

Hugh's understanding of the process of restoration flows from his belief that Christ, the Light-Wisdom, is at work within human beings, restoring the divine image that has been defaced by human sinfulness through their study of the arts, sciences, and Scripture. In the *magister* of St. Victor's epistemology, this restoration takes place on two levels within the knowing process: the world of corporeal objects and the world of spiritual objects. One moves from knowledge of the external world to knowledge of the internal world. The entire process of restoration is coordinated on all levels of knowing by the indwelling of Christ, the Light-Wisdom. The special illuminating function of Light-Wisdom is related to the Hugonian teaching on the three eyes: ''There is, however, a threefold eye: the eye of flesh, the eye of reason and the eye of contemplation. . . . By the eye of flesh we see those things which are outside ourselves; by the eye of reason those things which are within ourselves; by the eye of contemplation those things which are within and beyond ourselves'' (Hugh, PL 175:976a).

The activity of the divine Light-Wisdom within history restores vision to all three eyes, blinded by sin, through the healing power of the sacraments and the contemplative experiences of Christian life. Hugh says clearly: "Christ is both Word and Wisdom. . . . And Wisdom itself is light and God is Light because God is Wisdom. And when God illumines, God does not illumine by another light, but by that Light which God is because God is Light and the Word is Light and Wisdom is Light" (Hugh, PL 176:848).

Christ, then, for Hugh is clearly a dynamic reality, guiding the process of restoration as it touches every area of human activity: "Of all human acts or pursuits, then, governed as these are by Wisdom, the end and the intention ought to regard either the restoring of our nature's integrity, or the relieving of those weaknesses to which our present life lies subject" (Hugh 1961, 51–52). The illuminating function of the Word-Wisdom flows from the healing power of his humanity: "Truly the humanity of the Saviour has become medicine so that the blind may receive light and seeing by the light of doctrine they may recognize the truth" (Hugh, PL 175:926).

Christ the Divine Physician and the Sacraments

Christ, the divine physician, seeks to restore light and vision to fallen humankind through the healing power of the sacraments: "God has furnished [sacraments] at different times and places for humanity's healing: some before the law; others under the law, others under grace, diverse, indeed, in species yet having the one effect and producing the one health" (Hugh 1951, 150). Christ, the physician, brings healing to fallen men and women through sacraments which vary at different stages in salvation history.

> In this life itself and on account of it were the sacraments instituted, and some from the beginning of it, which ran in their own time and availed to restore health, in so far as it had been granted to them and was to be granted through them. And these sacraments, when their time was completed, ceased and others succeeded in their place to produce the same health. Again after these others were added, as it were, last of all, to which others were not to succeed, in as much as they were perfect medicines, so to speak, which would destroy the sickness itself and restore perfect health. And all these things were done according to the judgment and dispensation of the Physician, who saw sickness itself and knew what kind of remedies should be applied to it on every occasion. (Ibid.)

Hugh defines a sacrament as "a corporeal or material element set before the senses without, representing by similitude and signifying by institution and containing by sanctification some invisible and spiritual grace" (ibid., 154–55). Since matter had participated in humanity's fall, matter should play an integral part in human restoration. In the view of the *magister* of St. Victor, men and women are not meant to seek salvation from the sacraments, but in the sacraments from God.

The sacraments of the natural law, such as tithes, sacrifices, and oblations, and the sacraments of the written law, such as circumcision, confer their salutary effects, but not by properly communicating any power of grace. Rather, the effects are conferred by symbolizing the sacraments of the postincarnation period, which do contain the efficacious power of sanctification, flowing from the passion of Christ. The divine initiative for the healing of sinful men and women, concretized in the sacraments, asks for a faith response that will restore hearts and minds unto that contemplative vision which is the goal of creation.

Faith and Contemplation

Hugh applies his basic sacramental understanding to the relationship between faith and contemplation: faith is the sacrament of future contemplation. He then relates faith as sacrament in a general sense to Eucharist as sacrament in a specific sense: "Faith, then, is the sacrament (*sacramentum*) of future contemplation, and contemplation itself is the thing (*res*) and virtue of the sacrament, and we now receive meanwhile the sacrament of sanctification that sanctified perfectly we may be able to take the thing itself" (ibid., 181). The Christian receives the healing presence of the Lord in the Eucharist so that, through the sanctifying *res* of that sacrament, received in faith, he or she may be gradually restored to the *res* of contemplation: "the sacrament of faith is a true sign, but at the same time also essentially a grace-mediating symbol which from this power always leads to the vision of God and its blessedness" (Weisweiler 1957, 442).

Although the eye of contemplation within each believer is initially closed to the full vision of the Lord's presence because of original sin, the Lord, by the gift of faith, begins to remove that blindness so that one can begin, in some sense, to see again.

> If, then, the highest human good is rightly believed to be contemplation of the Creator, not unfittingly is faith, through which one begins in some manner to see the absent, said to be the beginning of good and the first step in restoration; this restoration, indeed, increases according to increases of faith, while the human person is enlightened more through knowledge that he or she may know more fully, and is enflamed with love that he or she may love more ardently. (Hugh 1951, 181)

Restoration to full vision begins in the darkness of faith, but that very faith is a sacrament which promises further restoration to the face-to-face vision of God. The risen Lord, illuminator and physician, who dwells within believers spiritually through the *res* of the Eucharist, seeks to draw them back by faith and love to the fullness of contemplation.

The purpose of the coming of Christ, the Light-Wisdom, in the Eucharist is not only to heal the believer's blindness unto contemplation, but also to initiate his or her vision of God amidst the darkness of faith. This new seeing in faith through incorporation into the saving work of Christ opens the possibility for a new mode of life to the believer: "When the person is lifted to contemplation . . . he or she sees as from a distance what one may call a land of light and a new country, such as one does not remember ever having seen before, and never thought existed" (Hugh 1962, 105–6).

At the final stage of the process of restoration the *res* of the Eucharist finds its fulfillment in the *res* of contemplation. Initially, the spiritual presence of the Lord shared with the individual through the sacrament is known only in the darkness of faith. But, by the continuous healing activity of the indwelling Spirit, poured into the mind and heart of the person by the Eucharistic Lord in each reception of the sacrament, the Christian begins by faith and love to move towards a new life.

As the contemplative ascent unfolds, the eye of contemplation is fully reopened and faith yields to a new understanding, not only of God, but also of all the works of God in creation. Hugh says, "The one occupation is to contemplate God's wonders and to praise God's works" (ibid., 178). In the contemplative experience that flows from faith and participation in the sacramental life of the Church, one receives new insight into the sacramental dimension of all created reality.

One's own experience of this new vision leads one to a deeper life of love and service within the Christian and human communities. For Hugh envisions not only individuals struggling to restore the image of God within themselves through faith and the grace of the sacraments, but every Christian community becoming a contemplative community, dedicated to new seeing and new acting as a community of restoration. By reason of common participation in the Eucharist which unifies them in Christ, the members of a community are responsible for assisting one another in the common work of restoration. The indwelling risen Lord, who illumines and heals them by the Spirit, impels them as members of a Christian community to a life of service, to assist one another and all the members of the human community in attaining the contemplative vision of God and God's creation to which they are all called.

Conclusion

Although the twelfth century, with its strong Platonic and Pseudo-Dionysian perspectives, was able to preserve a balance between the natural world as an object of rational inquiry and as revelatory of the mystery of God, the rediscovery of Aristotle and the growing ascendency of Aristotelianism, from the thirteenth century onward, tended to desacramentalize nature. Hugh of St. Victor represents a Christian presentation of the significance of the natural world in which the Aristotelian view has not yet made nature merely the object of scientific investigation.

The Hugonian vision of the sacramental dimension of the natural world can offer an important perspective for recovering a Christian understanding of the relationship between the natural world and human development. For there can be no truly human development that does not recognize and preserve the necessary interconnectedness of the environment and the human community. Hugh's belief that the elements of the natural world, already sacramental through the act of creation, assume a new dimension of meaning in becoming bearer's of God's grace in the sacramental economy established by Christ offers us a very significant insight for making a Christian response to the ecological challenge. It means that we must take seriously the grace-bearing dimension of all created reality.

Hugh's teaching that the *res*, or special gift of grace given in the Eucharist, opens the eye of contemplation and involves a new seeing that enables believers to encounter God, themselves, and the world around them "as a new country, such as one does not remember ever having seen before" (Hugh 1961, 105–6) provides an important link between the sacramental dimension of Christian experience and contemplation in action that is meant to flow from a community's celebration of the Eucharist. A community that has known the grace of God, mediated through the sanctified gifts of the earth in the liturgy, should also be a community committed to seeing anew the sacramental dimension of the natural world and to preserving the environment as the great sacrament of God's love, manifest in creation and renewed in the new creation that flows from the incarnation and redemption. As we struggle with the ecological crisis in the last decade of the twentieth century, perhaps we can, indeed, find some new insights by looking more closely at the experiential wisdom of our twelfth-century forebears such as Hugh of St. Victor.

REFERENCES

Chenu, M. D.
1968 *Nature, Man and Society in the Twelfth Century.* Trans. Jerome Taylor and Lester K. Little. Chicago: University of Chicago Press.

Hugh of St. Victor
1961 *Didascalion: A Medieval Guide to the Arts.* Translated from the Latin with an Introduction and Notes by Jerome Taylor. New York: Columbia University Press.
1951 *On the Sacraments of the Christian Faith.* Trans. Roy J. Deferrari. Cambridge, Mass.: Medieval Academy of America.
1879 *Opera omnia Hugonis de S. Victore. Patrologiae Cursus Completus Series Latina* 175–76. Paris: Migne.
1961 *Selected Spiritual Writings.* Translated by a Religious of C.S.M.V. New York: Harper & Row.

Ladner, Gerhart B.
1982 "Terms and Ideas of Renewal." In *Renaissance and Renewal in the Twelfth Century.* Eds. R. L. Benson and Giles Constable. Toronto: University of Toronto Press.

Southern, R.W.
1971 "Presidential Address: Aspects of the European Tradition of Historical Writings: Hugh of St. Victor and the Idea of Historical Development." *Transactions of the Royal Historical Society* XXI (1971) 159–79.

Weisweiler, Heinrich
1957 "Sacramentum Fidei: Augustinische und ps. dionysiche Gendaken in der Glaubensauffassung Hugos von St. Viktor." In *Theologie in Geschichte und Gegenwart,* eds. J. Auer and H. Volk, 433–56. Munich.

9

The Earth in American Catholic Spirituality

John Manuel Lozano, C.M.F.

A Short Prelude

Catholicism, like Eastern Orthodoxy, bears the deep imprint of the Mediterranean shores on which it grew up. Like the many little white chapels of Italy or Greece that open out upon luminous vistas, Catholic religiosity remains in continuous contact with the earth and its products: the water of baptism, the wine and bread of the Eucharist, the oil of anointing, the wax of candles, the roses from the garden. The norms for the liturgy insist on all of these things: the wine must be entirely from grapes, the bread from wheat, the oil from the olive, and the wax from the beehive. These symbols are commonly Christian, but they play a much larger role in the rural religiosity of Catholicism or Orthodoxy than in the urban spirituality of Protestantism. This predominance of gestures in Catholic liturgy may well reflect the influence of environment. For, whereas the word may flow free of gesture in the European north, the gesture hardly ever abandons the word in the south. It is a well-known fact: gesture is but a word made visible. And sight, smell, and touch count a great deal around the Mediterranean, a region rich in forms and colors. Indeed, for a peasant, they count even more than hearing.

While a good number of the first Christian communities were urban (in the fourth or fifth centuries non-Christians began to be called *pagani,* marsh-dwellers, peasants, villagers), they were communities heavily influenced by the agricultural hinterland. Athens and Rome always looked to the fields. This helps us form a better understanding of the presence of these fruits of the field in the liturgy. It also helps us appreciate the feeling for the body in Catholic and Orthodox spirituality. Processions, medieval prostrations before statues (every night, we are told, Saint

Dominic used to rush about his church making them), the cultus of the bodies of the saints, and even the use of hairshirts and disciplines, are all reflections of the importance of the body in rural cultures. And it is significant that all of them began to be criticized by the city folk of the devotio moderna, before they were rejected by the Reformation and relegated (as far as possible) to the sidelines by the Jesuits (an eminently urban order). Not so long ago, while I was attending a worship service in a rich Presbyterian church, I was reminded of my contacts with the Jesuits in my childhood days: the same black robes, the same immobile bodies, the same inwardness, with the word floating over the assembly and seeping into their receptive minds.

This happened to all of us, some four centuries back, in Europe. We exchanged the discipline and the hairshirt for interior mortification, and meditation came to predominate over the liturgical gesture. Then North Americans discovered jogging, diets, and health food (that is to say, the body), the Saint Louis Jesuits turned their hand to the guitar, the charismatics began hugging one another, and young nuns rediscovered the liturgical dance.

When we speak of the experience of the earth among North American Catholics, we have to begin with starting points outside liturgy. We will try to be specific about these. But one thing is certain: although by historical pathways distinct from those of their European counterparts, in the religious experience of American Catholics, as well as in other religious traditions of the United States, "the earth and all it contains" has played an important role.

The Influence of Myth

Let us begin with the influence of the common myth of origins. The United States is the only modern nation that celebrates its origins with a myth and a household rite: the account of the arrival of the first Protestant settlers in what would come to be New England, and the family meal that commemorates and celebrates that event. Exactly like the biblical people of ancient Israel, Americans, together with the Canadians who imitated them, are the only people who celebrate a feast of Thanksgiving, which originally expressed gratitude for the fruits of Mother Earth throughout the year.

Clearly, this story has assumed the proportions of a myth of origins. In it Americans have summed up some of the fundamental values of their society: the religious meaning of life, the search for liberty and happiness, solidarity with other ethnic groups. . . . It is just as clear, however, that the myth bears the imprint of the white Protestant culture in which it was shaped. The Protestant Pilgrims who arrived in Massachusetts were the first to emigrate to these shores in search of religious freedom, but they were followed a little more than a decade later by the Catholics of Maryland. Florida, discovered by the Spaniard Juan Ponce de León in 1513 and subsequently explored by Pánfilo de Narváez and Hernando de Soto, saw the founding of the first major settlement in St. Augustine in 1565 and, thanks to the missionary efforts of the Franciscans, could claim some thirty thousand Indian converts before the disastrous British occupation of 1763–83. While we commemorate the Mayflower, we tend to forget those other ships that came here, crammed with shackled African slaves who, though not "Pilgrims," also form part of our roots. Doubtless, too, the Wampanoag and other Amerindian groups held a quite different view of the arrival of these various waves of European colonists. Perhaps their presence in the myth, offering us the fruits of the earth, serves in a Freudian way to exorcise us of our guilt complex for what we have in fact done to them.

Catholics eventually adopted the holiday, the commemorative myth, and the accompanying ritual celebrating it. John Ireland, archbishop of Saint Paul, himself an immigrant and an enthusiastic neo-American, took great pride in proclaiming the national festival.

Indeed, the commemoration became so popular largely because it struck a chord of common experience among the teeming masses of immigrants. The fruits of the earth symbolized freedom from the daily hunger that had oppressed most Europeans. I still remember the emotion with which a Sicilian-born gentleman spoke to me of the fruits grown in various states of the Union, which were, according to him, incomparably better than those of his native Italy! And although he had come to savor them by dint of hard labor, he had done so free of the ancient miseries of his earlier years.

God's Country

In fact, even more than the fruits of the earth, it was the earth itself, the land, that immigrants experienced as a gift of God. John Ireland glowingly portrayed the North American continent as "the fair region of His predilection, which He has held in reserve, awaiting the propitious moment in humanity's evolution to bestow it on men, when men were worthy to possess it" (Ireland 1896, 139). While inviting Catholics to head west to Minnesota (and taking a polemical jab at the archbishop of New York, who wanted to keep them there), John Ireland exclaimed: "Man made the city but God made the country!"

For many, indeed for all who crossed the Appalachians and the Mississippi, America was God's country, a godly land that had been given to them so that they could build for themselves an existence worthy of the sons and daughters of God. Perhaps even more than the wild mountains they had crossed, it was the vast midwestern plains that filled them with a sense of wonder. Their dreams of having a plot of land to call their own began taking shape. It is curious to note that among Catholic immigrants, most of the Irish and nearly all of the southern Italians fled the land and holed up in cities, near the great factories. The earth held too many bad memories for them. It was the Germans, many of them Catholics, who settled along both sides of the Mississippi from Saint Paul to below Saint Louis.

Since many of these immigrants were illiterate, they were unable to leave a written record of their feelings. But by dint of repeating to their children the story of their almost religious love for their new American homeland, they have made it part of our life-blood.

Through Mountains and Valleys

With, or in the footsteps of, the immigrants came a number of women who not only knew how to write but also had to recount the vagaries of their travels to their sisters in Europe. Thus was born a Catholic travel literature to which the Frenchwomen Philippine Duchesne and Théodore Guérin and the Italian Frances Cabrini were eminent contributors. They allow us to glimpse how the new American land unfolded before the eyes of immigrants as they made their halting way by carriage or boat.

Rose Philippine Duchesne (1769–1852), after crossing the Atlantic from France, arrived in New Orleans as leader of the first band of the Sisters of the Sacred Heart. From New Orleans she traveled up-river by boat to Saint Louis. From the deck Philippine took note of the varied panorama that lay on either side of the "Father of Rivers."

Théodore Guérin (1798–1856) also arrived from France, but entered the continent through the port of New York. She crossed the Alleghenies in a creaking old coach. As she went, she let the wild beauty of the scene pass through her eyes into her heart, discerning behind it the Author of it all. "At every turn new grandeurs rose before us. Sometimes we were on heights where mountain-tops were our footstools; below were superb defiles where magnificent valleys spread their verdure. . . . The eye is lost in the ravishing spectacle, so calculated to elevate the soul towards the Author of all things" (Guérin 1978, 441).

But of all these well-travelled women, the best-travelled by far was the one from Northern Italy, Frances Xavier Cabrini (1850–1917), foundress of the Missionary Sisters of the Heart of Jesus. Frances, following the route of so many of her compatriots, arrived in New York for the first time in March of 1889. She would cross the Atlantic several times, between France or Italy and New York, or between Buenos Aires and

Cadiz, en route to Barcelona and Genoa. She sailed from New Orleans to Central America, going overland to continue along the Pacific coastline to Chile, until at length she boldly crossed the Andes into Argentina. She traveled through North America in several directions, from New York to Denver and Seattle and from Denver along the Colorado River to Texas, then turning back to Louisiana Frances Cabrini was touched to the heart by the beauties of nature and enjoyed describing them to her Daughters.

As she saw it, Seattle was the garden place of the United States. "While the ever snow-covered peaks of the Rocky Mountains on the one side and the Olympics on the other brought us far-off reminders of the Pole, the verdant hills, touched by the tides, were aflower with orange and lemon blossoms, and seemed rich in splendid vegetation." (My translation; cf. Cabrini 1955, 241).

Naturally, Frances had been deeply touched by esthetic and religious sentiments at the imposing heights of the Rocky Mountain:

> The mountains are immense masses of rock, colored with the most beautiful tints of the rainbow. They present a most enchanting view, and form one of the great natural beauties of the United States. If one were to see this scene painted, those enormous masses that seem to hang by a thread, with the railway cars running zig-zag between the folds of the mountains up to the highest peaks, and then precipitating themselves down into the valleys below, and running through the gorges called cañons, whose walls are inaccessible and, because of their marble-like colours and beautiful forms seem like an enchanted castle, one would imagine the whole thing was simply a creation of the painter's brush. (Ibid., 233)

She has left us several written snapshots of her trip from Denver to New Orleans. On the Grand Canyon: "For many hours the train runs through very narrow gorges called Canyons, famed the world over. They form two inaccessibly high vertical walls that seem to touch the sky, while down below the ever winding river, sometimes calm, sometimes agitated, reflects in its waters the varied hues of the most marvelous rocks I have ever seen" (cf. ibid., 247). Then, as she entered the flatlands of Texas: "The immense plains of Texas, for the most part uninhabited because of the vast expanse of this State, are most fertile and suitable for all kinds of cultivation. Still virgin lands, rose-colored, full of life and promise" (cf. ibid.).

God in a Chestnut Tree

Nature-love had not begun with the numerous waves of immigrants who arrived during the nineteenth century. Nor were they the first Americans for whom nature was the setting for an authentic religious experience.

Long before the arrival of these women born in France or Italy, all of this had been felt by a young upper-class girl, born in one of the East Coast Colonies shortly before the War of Independence (1774) to a family that belonged to the Church of England. Her name was Elizabeth Ann Bayley or, as she was known after her marriage, Elizabeth Ann Bayley Seton. If we mention her here, and in a prominent place, it is because when she later joined the Catholic Church, Elizabeth became one of the Mothers of our Church.

Elizabeth had been marked out by suffering at an early age; she was still a little girl when her mother died. Her father, Dr. Bayley, had to surround her with a great deal of affection. Even this made her vulnerable. What would happen if her father were to die or grow tired of her? God intervened to unite her with the Love that never fails. It happened while her father was on a voyage to England in 1789, when she was barely fifteen years old and in the full blush of her physical and emotional development. Years later, she herself gives us an account of this special experience:

> In the year, then 1789, when my father was in England, one morning in May I set off to the woods about a mile from home, and soon found an outlet to a meadow, and a chestnut tree attracted my attention, and when I came to it I found rich moss at the foot. There, then, was a soft seat; the sun was warm, the air still, and a clear blue vault above; and all around I heard the numberless sounds of the joy and melody of spring. The sweet clover and wild flowers I had gathered by the way were in my hand. I was filled with love of God and admiration, enthusiastic even of His works. I can still recall many sensations that my soul felt at the moment. I thought my father did not care for me—well, God was my Father, my all. I prayed, sang hymns, cried, laughed and talked to myself of how far He could place me above all sorrow. There I lay still to enjoy the heavenly peace that came over my soul, and I am sure I grew, in the two hours so passed, ten years in my spiritual life. Well, all of this came vividly to my mind this morning when, as I tell you, the body left the spirit alone. I had prayed and cried heartily (which is my daily and often hourly comfort), and closing my eyes with my head resting upon my arm on the table, lived all those sweet hours over again, made myself believe I was again under the spreading tree—felt so peaceful a heart, so full of love to God, such confidence and hope in Him. (Seton 1869, 1:121)

The experience that Mother Seton narrated in 1803, some fourteen years after the event, had a fundamental spiritual and psychological value. Looking back at it over the years, she was convinced that during those two hours she had grown ten years in her spiritual life. It left such a deep impression on her that she relived it fourteen years later amid extraordinary circumstances: shut up in a *lazaretto,* with her husband dying and her children around her, a few weeks before Catholic piety would touch her so deeply that she decided to enter the Church of Rome. It was an instance of one of those revelatory experiences that orient a human being toward a particular type of spirituality. When they occur, as they did here, during childhood or adolescence, I prefer to call them "pedagogical graces." From that moment on, Elizabeth lived in the presence of her Father in heaven. That revelatory and orientating experience took place in a natural setting of such beauty that it seemed that everything about it had been prepared just for her: the flowers she had picked, the meadow in the woods, the warm sun, the still air, and the blue vault above . . . and she had sat down upon the rich moss. The active contemplation of an ascetical type here gave way to a mystical intuition and delight of great immediacy. Thanks to it all, Elizabeth had encountered her real Father forever. From then on, her spirituality would be stamped by a filial confidence and tenderness.

Adam Returns to His Mother

So a great woman, little more than a child at the time, found her Father while reclining on the bosom of her Mother Earth. Yet her eyes were fixed only on her heavenly Father. But where was mama? Ah, but as you know, dear Electra, there is a time when young girls only have eyes for their daddy. Much later, two great American men, both of them, like Elizabeth, converts to Catholicism, would also sense the hug of Mother Earth.

The first of them was Isaac Thomas Hecker, a man endowed with a strong love of Nature that was quite romantic and quite American:

> I love to put my naked feet upon the bosom of the Earth and feel the gentle breezes play about my body.
>
> The Earth heaves and sighs from its very heart in sympathy for Man's woes and sorrows, and Man rests upon her bosom as upon the bosom of a kind mother, and she drinks up his bitter tears in compassion, and extracts the painful poison from his heart, and pours in his heart instead the waters of joy and gladness.
>
> We are Sisters to Nature: She is a child of God as we are, and she has partaken of the same penalty of Sin that we have.
>
> Nature is redeemed by the same Redeemer that we are and by no other. (Hecker 1988, 196)

It was a return to Genesis. Adam had gone back home to Mother, a tired, sorrowing man looking for rest and peace. Hecker often liked to refer to that "paradisiacal" human being to whom we must submit our history of sin.

Yes, the Earth is Mother, but Nature is Sister to this man who had also accepted his *anima,* his own feminine side. And this he probably owed to his close relationship with Mother Earth.

It was not an isolated instance. Hecker liked to retreat into the "silent woods," the "sanctuaries of God." There, he repeatedly met his Mother:

> This afternoon I took my Greek books and Shakespeare's Sonnets and went to the woods.
>
> Lying on the bosom of my Mother looking up into the heavens of my Father while the birds were singing in the Trees. The gentle winds filling the space with the sweet fragrance of spring flowers. The water flowing so slowly along mirroring the whole firmament above. Oh I feel like being in my native home: This is Eden. (Ibid., 175)

If he loved the earth with a son's heart, he loved all the creatures that move on her surface. One day, he could not resign himself to the idea that animals die forever, and he dreamt of a life-after-death for them: "Who can look upon the animal creation in all their divine beauty and deny them a further existence. See their graceful motions! their beautiful forms! and the many pure instincts they exhibit! Did they not share in common with the curse that was pronounced upon Man? And are they not promised a participation in the restoration, in the Millennium?" (ibid., 278–79).

The second great man was Thomas Merton. He, too, liked to walk through the woods around the monastery and he felt strengthened by his contact with Mother Earth:

> Heavenliness again. For instance, walking up into the woods yesterday afternoon, it was as if my feet acquired a heavenly lightness from contact with the earth of the path; as though the earth itself were filled with an indescribable spirituality and lightness. As if the true nature of the earth were to be heavenly; or rather, as if all things in truth had a heavenly existence. As if existence itself were heavenliness. One sees the same thing obviously at Mass but here with a new earthly and yet pure heavenliness of bread. (Merton 1988b, 44)

Merton's experience is the same as Hecker's though the images and vocabulary are different. It may be, too, that in his younger years, Merton may have been looking for his dead father in heaven. By this time, however, he had read too much Sufism and too much Zen. There was only Mother Earth. But Mother, as in a Zen experience or in a Japanese painting, had become pure heavenliness. She had become the screen of a religious experience in which earth appeared permeated by God's presence.

Sometimes he felt a kind of Zen experience when listening to the song of a bird or watching the sudden graceful apparition of a deer near his hermitage: "The deer reveals to me something essential, not only in itself, but in myself. . . . Something profound. The face of that which is both in the deer and in myself" (ibid., 208). And his sense of communion with the universe was strengthened: "One thing the hermitage is making me see is that the universe is my home and that I am nothing if not part of it. Destruction of the self seems to stand outside the universe. Get free from the illusion of solipsism" (ibid., 156).

The Earth: from Shapes and Color to Pain

Thomas Merton inherited from his parents both their devotion to the landscape and a painter's eyes to describe it. Or perhaps to narrate it; for, like the impressionists, Merton knew the ephemeral nature of a scene, the rapidly changing rhythm of its lights, colors, and shades.

He repeated his story many times. One day during his student days in England, he sat silently in the country, watching the wheatfields strewn with poppies, and the birds. After a while he left, but as soon as he turned his back he felt the need to come back to the same spot. He explains why: "That was because I knew that when you have seen a place once, you would never see it exactly like that again; the light would be different, and the air and sky and shadows and colours would be different, and living things would have grown into different shapes, and many old things would have perished while new things would have grown into their places" (Mott 1984, 58).

Later, Merton developed a religious sense before what he saw. In his final years, when he had already learned the Zen art of unity and depth, his heart was no longer "raised" to God by the beauty of mountains and canyons, as had happened centuries earlier to Augustine, and only decades earlier to Mother Guérin or Mother Cabrini. His contemplation of trees, creeks, and deer was in itself a religious experience. He did not need to mention God. God became transparent in a frail creature.

By his own decision, Merton remained confined for most of his life to the same horizons: the rolling hills of Kentucky and the woods around Gethsemani. We have already seen how deeply his experience of the earth and his spirituality were interwoven. Towards the end of his life he began to travel again. His journals could turn from the interior world of his soul and his monastery and the books he was reading to the country he was discovering. His journals of the American West (California, New Mexico, Arizona, Alaska) are a mixture of pictures and thoughts (the ocean and the Astavakra Gita, the mountains and Unamuno, the forests and Sartre), while in his Asian journal, Asia is more present as a culture and a religion than as a geography. The airplane, rapidly, unrolls the land below: "Blue shadowed mountains and woods under the cloud, then tiny shining, tin roofed houses at a crossroad. An olive-green valley floor. A low ridge thinly picked out at the very top in blown snow. The rest deep green. One of the most lovely calligraphies I have ever seen" (Merton 1982,

8). Another time it was the Alaskan mountains: "Most impressive mountains I have seen in Alaska: Drum and Wrangell and the third great massive one whose name I forget, rising out of the vast birchy plain of Copper Valley. They are sacred and majestic mountains, ominous, enormous, nobly stirring. You want to attend to them. I could not keep my eyes off them" (Merton 1988a, 20).

It was not only or mainly the grandiose that attracted Merton. There was also a small fishing town: "A small fishing town between steep mountains and blue water—a highway on one side, and Eyak Lake around at the back" (ibid., 19). And the flowers at Bear Harbour: "Besides wild irises three or four feet high, there are calla lilies growing wild among the ferns on the stream bank. A profusion of roses and a lot of flowering shrubs that I cannot name" (Merton 1982, 16). In the New Mexican desert, "A gang of gray jays flied down into the canyon with plaintive catlike cries over my head" (ibid., 30). Shapes, colours and sounds are fixed forever.

Thomas Merton was also one of the first Americans to become aware of the tragedy of the earth being destroyed by human greed. He discovered the damage that had been done to the redwoods: "The redwood lands appear. Even from the air you can see that the trees are huge. And from the air too, you can see where the hillsides have been slashed into, ravaged, sacked, stripped, eroded with no hope of regrowth of these marvelous trees" (ibid., 11). Later, driving down through the redwoods, he discovered "one long stretch where the big trees have been protected and saved—like a completely primeval forest" (ibid., 12). Even though he says nothing more (he was very reserved about his personal feelings), he must have felt deeply hurt. The destruction of Mother Earth was becoming all too evident.

The Earth between Courtly Theology and Mystical Communion

Merton's final attitude towards the Earth and all that fills it is an exemplary case of what has been going on in Christians' attitudes towards the universe. The Fathers and Doctors of the Church were fond of quoting Genesis 2:19-20 about the primordial human being assigning names to each of the animals—an action in which they saw the royal eminence of humankind. In more recent times, we were more prone to recall Genesis 1:28-30: "Rule the fish in the sea, the birds of heaven, and every living thing that moves upon the earth . . . I give you all plants." We used this text to justify our gradual destruction of the planet.

But if there is a courtly exegesis and theology, ready to justify human privileges, there has also been a protest of the Spirit through the mystics. Francis of Assisi felt a deep sense of communion with Brother Sun and

Sister Moon. John of the Cross compared the Beloved (the Son of God, image of the ultimate Source of all) to mountains and valleys and even to the exotic beauty of the far distant isles of the newly-discovered America (Canticle 14.8). In our days Thomas Merton has joined them, experiencing how the earth is permeated by the divine to the point of becoming heavenly, how a quietly approaching deer reveals something profound: the common bond between animals and human beings (cf. Merton 1988b, 208). The pain he felt over the destruction of the redwood forest was nothing less than a logical consequence of his mystical sense of communion.

From mystical communion to prophecy.

In his last years, especially during his hermitage days, Merton joined the list of Catholic mystics who had become prophets. His proclamation and denunciation were centered on the search for peace, the promotion of civil rights, and the popularization of an ecumenical mind and heart in his Church. The defense of the earth had not yet become one of the priorities of Merton, the Berrigans, and their friends.

It was not until the 1980s that ecology became an important and urgent topic for American Catholics, joining other Americans to defend the environment. Proclaiming both the need for peace and the defense of the habitat, a motherhouse of sisters was proclaimed by themselves a nuclear-free territory. The message given by the purely symbolic gesture was heard and one well-known conservative poured bitter irony over the sisters.

The earth has become an integrated part of American spirituality. Motherhouses of sisters (in Lagrange, Illinois, for example) have begun to celebrate the day of Mother Earth. Here ecology and feminism blend. Women are perceiving a subtle connection between the subjection of Mother Earth to a pure logic of power and profit and their being made objects of male dominion and desire. "Ecology and spirituality" has become a rather common topic in Catholic seminars and workshops.

Small groups of Catholics have joined others in celebrating the passing of the equinoxes with rituals borrowed from native Americans, and celebrating nature with its miraculous energy. Of course, some times it is forgotten that Easter was initially the celebration of the spring (the new-born lambs) and Christmas was instituted to Christianize an old Roman feast acclaiming the rebirth of the sun.

Clearly, then, the earth has played a substantive role in American Catholic spirituality. The bases laid by the early pioneers as well as the more recent writers have given us a spiritual legacy which can serve us well in the critical years ahead as we face the challenging task of preserving our planet.

REFERENCES

Cabrini, Frances Xavier

1955 *Travels of Mother Frances Xavier Cabrini.* Milwaukee: Cuneo Press.

Guérin, Théodore

1978 *Journals and Letters.* Saint Mary of the Woods, Ind.: Providence Press.

Hecker, Isaac Thomas

1988 *The Diary.* New York: Paulist.

Ireland, John

1896 *The Church and Modern Society.* Chicago: D. H. McBride.

Merton, Thomas

1982 *Woods, Shore, Desert.* Santa Fe: Museum of New Mexico Press.

1988a *The Alaskan Journal.* Isla Vista: Turkey Press.

1988b *A Vow of Conversation.* New York: Farrar-Strauss-Giroux.

Mott, M.

1984 *The Seven Mountains of Thomas Merton.* Boston: Houghton Mifflin Co.

Seton, Elizabeth

1869 "Memoir." In *Letters and Journal of Elizabeth Seton.* New York.

Part V

An Interreligious View

10

"Do Not Destroy"—Ecology in the Fabric of Judaism

Rabbi Hayim G. Perelmuter

Two Childhood Memories

There was a "blue box" in our kitchen. It had a map of the land of Israel on it, and we regularly dropped our pennies in it. It was to purchase and redeem land through the Jewish National Fund. The goal: to transform desert into productive land, to cover denuded hills with trees. I did not know at the time that ecology was involved.

The other memory is of the visit of a leader of the Jewish National Fund in what was then Palestine to our home. I was twelve at the time. My father and he discussed the process of contacting leaders of the Canadian Zionist Movement. The goal: to win support for the project to raise $100,000 a year for ten years to purchase a huge swamp area in the middle of the country, to be drained and made useful for agriculture. I did not realize it then, but it was another lesson in ecology for me.

Forty-five years later, in the summer of 1971, I found myself in that area at a seaside dune, right at the edge of where this swamp had been, hardly twenty kilometers north of Tel Aviv. I had the occasion to see Kfar Vitkin and the Hepher Valley area, now lush, fertile, and productive, veritably the breadbasket of Israel. In the town square was a plaque recording the fact that this valley had been reclaimed by Canadian Jewry. And it all started with those conversations when I was just a stripling. The lush fields and farms where once mosquitoes and malaria held sway, awoke those memories. Ecology indeed. Redemption of land from desert, swamp, and desolation.

This aspect of redemption, redemption of fertile and productive land from arid desert and wasteland, which came into my own childhood consciousness in the development of the movement for the national rebirth of the Jewish people was deeply influenced, I think it fair to say, by deep currents within Judaism that had to do with accepting this world, loving it, and taking seriously its stewardship. Looked at this way, this redemption is perhaps the heart of what we see as the ecological goal.

Ecology in the Jewish Tradition

It is our purpose to examine the motif of ecology at two levels, first in the biblical and rabbinic approach, and second in what the cycle of the Jewish festival year can tell us about ecology and its central role in Judaism.

The Hebrew Bible gets a mixed review from ecologists. There are those, like Lynn White, who argue that human arrogance towards nature derives ultimately from attitudes already present in the Bible's creation stories (White 1967, 1203–7). The "rule over it" of Genesis 1:29 is taken for a mandate to justify the policy of "slash and burn" and devil take the hindmost.

Arnold Toynbee echoes this view. He reads the "rule over it" verse of Genesis as a "license and an incentive for mechanization and pollution" (Toynbee, 1971). And if one were to believe landscape architect Ian L. McHage, it was the sanction and injunction "to conquer nature, the enemy of Jehovah" (Harvard 1991, 8).

There is a long and sorry tradition of this kind of view. It is part and parcel of the approach that tended, almost on reflex, to negate Judaism and its biblical roots. Here, from the hindsight of triumphalist Christianity with its supersessionist motifs, "Old Testament" Judaism was seen as vengeful, legalistic, under a somber divine authority, in which humankind, in the image of a vengeful, angry God, was similarly vengeful and angry in relationship to nature. Robert Hamerton-Kelly's description of biblical Judaism as violence incarnate is a recent example of the extremes, wittingly or unwittingly, to which these views can be taken (Hamerton-Kelly 1992, 8). The earth and its produce was to be exploited by humankind. Such virtues as compassion, gentleness, caring, love of fellow human, and forgiveness were "New Testament virtues" foreign to the hard-hearted "Old Testament." Jewish law was mean, violent, nasty. Scholars like Wellhausen, Harnack, Bousset, and Bultmann, just to mention a few, nurtured this negative view.

The other view is getting its day in court. One has only to name George Foote Moore, E. P. Sanders, J. G. D. Dunn, Albert Schweitzer, W. D. Davies, Krister Stendahl, David Flusser, and Samuel Sandmel, to see the subsoil for that approach. More and more the perception that a careful reading of the Jewish Scriptures, particularly as read and exegeted by rabbinic sources that gave us Mishnah, Talmud, and Midrashim yields a totally different picture. It is a reading that enabled Bernhard Anderson to observe that "the biblical motif of human domination over nature, when understood in the full context of Israel's creation theology, calls into question the present practices of exploitation and summons people to a new responsibility" (Harvard 1989, 7). Theodore Hiebert (ibid., 9) could add: "Israel's world view, which did not labor under the dichotomies between nature and history, between matter and spirit . . . may once again provide some models and symbols for more integrative ways of understanding the relationship between human society and its environment."

It comes to this. Perhaps now is the time for the theologians of today once again to affirm a few simple truths: it was not we who created the world, it is not we who control it. It is our responsibility to learn to live with all other creatures, human and otherwise, within the limits of that world. What we should least wish to do is to complete the opening sentence "In the beginning God created the heaven and the earth" with "In the end we choked the heavens and destroyed the earth."

This becomes possible when it is recognized that, so far as Judaism is concerned, the essential meaning of the Bible is not simply on the surface of the text, but in how the compilers and editors read it and understood it. When we want to understand what the Bible really has to say on an issue we look at its texts as they are refracted through rabbinic commentary. The way in which the rabbis understood Scripture gives us a clue as to where the essence of the Jewish theological position is to be located.

What we know now as rabbinic Judaism, innovative, imaginative, and immensely creative, grew up within second commonwealth Judaism, and as though to "anticipate the cure before the disease," developed that crucial change which shaped a Judaism that could survive the destruction of the second Jewish state with a normative Judaism and an evolving Christianity (Perelmuter 1989, 5; also, Soncino 1938, 13b). It shaped Jewish life and Jewish tradition through the hermeneutic process of Torah interpretation and commentary. Here the attitude of Judaism can be discovered and decoded. It is to the rabbinic reading of Scripture that we turn for the Jewish position.

Let us begin with the already cited Genesis 1:28: "God blessed them and said to them: 'Be fertile and increase, fill the earth and master it [*u'r'du va*]; and rule the fish of the sea, the birds of the sky, and all the living things that creep on earth.' " What an indignant ecologist sees as a mandate to rule or ruin is understood quite differently in rabbinic commentary. The crux lies in the verb *r'du,* usually translated as "rule." Rabbinic commentary depends on a pun to interpret the passage. To be sure there is a Hebrew root *rada* that means "rule." But there is also a root *yarad* which means "to descend," and whose imperative form is also *r'du.*

Here the pun comes to the rescue of ecology. Both the Midrash to this passage and the Rashi commentary tell us: "If man is deserving, he rules; if not he is diminished." The ultimate test for humankind therefore comes in its attitude to ecology and stewardship.

Nor is this the only interpretation of *r'du* that is not obsessed with the idea of mastery. In the Talmud (Yebamot 65b) we read that in the context of this passage, "rule" suggests "procreation" and not "mastery." At bottom the rabbinic conviction is that the "earth is the Lord's" and that we are not so much its masters as its stewards.

From this point of view Genesis 1:28 is clearly associated with Genesis 2:18, where we learn that God "placed him [Adam] in the garden to till it and guard it." Clearly stewardship rather than mastery seems to be the guiding principle and supports the rabbi's pun.

Jeremy Cohen, in a thorough study of how Genesis 1:28 was read and interpreted in both Jewish and Christian commentaries, points out that the post-Reformation negative conclusion proceeded from a ". . . flawed methodology. Scholars simply assumed that their own understanding of the verse matched that of the author, that this understanding characterized Jewish and Christian readers of the Bible throughout the intervening centuries. . . . Hence it was perfectly permissible to link the verse directly to social and scientific tendencies of our own day and age" (Cohen, 1989).

Another important building block is to be found in Deuteronomy 20:19-20: "When in your war against a city you have to besiege it a long time in order to capture it, you must not destroy the trees wielding the axe against them. You may eat them but you must not cut them down." Indeed a crusading ecologist in Oregon would enthusiastically seize upon this verse. One is reminded of Astroff in Chekhov's *Uncle Vanya* who could say:

> Well, I admit you can cut woods out of some need, but why destroy them? Russian woods are creaking under the axe, milliards of trees perish, dwellings of beasts and birds are emptied, rivers go shallow and dry, wonderful landscapes vanish, never to be brought back again, and all because lazy man hasn't enough sense to bend down and pick up fuel from the ground. . . . He must be a reckless barbarian to burn this beauty in his stove, destroy what we cannot create again . . . (Chekhov 1929, 84)

The rabbis took this verse to represent a principle of ecological responsibility and concern. They extended the prohibition of cutting trees to all wanton acts of destruction (B. Kiddushin 32a). They extended the prohibition to matters of peacetime concern, including air pollution, restrictions on where tanneries could be located, and the placing of cemeteries near cities (Mishuah, Baba Batra 2:8-9; B. Baba Kama 82b). They extended its application from objects of nature to human artifacts. "Whoever breaks vessels, or tears garments or clogs up a fountain or does away with food in a destructive manner, violates the prohibition of *bal tashhit* [you may not destroy]" (Gordis 1986, 119–21).

Underlying this principle is the clear recognition that what we are wont to call our property is really God's. It is a misreading of Scripture to suggest that the Bible is person centered, that the human bestrides the world like a colossus and can do anything with nature. The "image of God" in us does not make us God. It does make us God's agent or steward, and God wants God's world hallowed and preserved. That is what *mitzvah* (divine commands to humans) is about, and doing ecology is a supreme *mitzvah!*

Even when it comes to animal sacrifice as a means of worshipping God, this principle comes through clearly. We learn in Leviticus 22:28 that "no animal from the herd or from the flock shall be slaughtered the same day with its young." The vegetarian would shrug his or her shoulders and ask, "Why not spare them all?" But when we read the commentary of Nachmanides on Leviticus 22:28 which says that "Scripture will not permit a destructive act that will bring about the extinction of a species, even though it has permitted the ritual slaughter of that animal for food," the ethic that translates itself into environmental awareness emerges clearly.

Samson Raphael Hirsch and Umberto Cassuto further corroborate this classical Jewish perspective. Hirsch, a great nineteenth-century theologian, insists that the prohibition of purposeless destruction of fruit trees around a besieged city is only to be taken as an example of general wastefulness. Under the concept of the principle of *bal tashhit* (do not destroy), the purposeless destruction of anything at all is taken to be forbidden, so that our text becomes the most comprehensive warning to human beings not to misuse the position which God has given them as masters of the world (Pelkovitz 1976, 102). And according to Cassuto, "legal cases recorded in the Torah . . . show important points in which Israel's conduct was to be different and superior to their contemporaries in this regard to emphasize the importance of the protection of the environment in a civilization where such considerations were not accepted" (Freudenstein 1970, 407).

Ecology in the Jewish Liturgical Tradition

Applying the rabbinic principle of scriptural hermeneutic to the cycle of festivals of the year yields similar results.

Let us begin with the Sabbath, the keystone of the arch of the festival cycle that shapes Jewish life. Of the Sabbath, the Hebrew thinker Ahad Ha'am once said, "The Sabbath has kept the Jew more than the Jew has kept the Sabbath" (Ahad Ha'am 1921, 79). It is the analogue to God's resting after the process of creation. It is a recurring day in a seven-day week, in which there is an opportunity to stand back, reflect, and achieve a taste of the messianic fulfillment at the end of time, to catch a glimpse of a world at peace. How significant that this command includes your "manservant and maidservant, your ox and your ass": it extends to all creatures of creation.

More than this, it extends to the environment itself. How else can we understand the Sabbath of the cycle of years, the sabbatical year, the command that every seventh year the land itself enjoy a Sabbath? The land is to lie fallow, to rest, to restore itself, to reinvigorate itself. An instinctive insight into how things ought to be. What a profound insight, in a prescientific era, of how what is taken from the land must be allowed to return to the land.

And what about the Sabbath of Sabbaths every seven years, that fiftieth year that is to recur to the end of time, the year to which we give the name Jubilee, when everything returns to its original owner, all bets are off, all slaves freed, all land returns to its original owner? Surely this is a signal to us, placed at the very heart of the religious cycle, that we are not masters of everything, that we are in fact stewards at best, stewards at risk in our stewardship if it is a flawed, destructive stewardship.

Abraham Joshua Heschel was right on target when he wrote of the spirit of the Sabbath, which he saw as the spirit of Judaism and vice versa: "There are three ways in which we may relate ourselves to the world—we may exploit it, we may enjoy it, we may accept it in awe." (Heschel 1955, 97). Accepting the world in awe is good advice in an approach to ecology, and it comes straight out of the spirit of Judaism.

Furthermore, Rosh Hashanah and Yom Kippur, the New Year and Day of Atonement, manifest an ecological concern in cutting the human species down to size. Hubris is taken to the woodshed and we spend much introspective time coming to understand that God rules, that we do not run the show, that we are not quite the masters of our fate that we think we are. We stand in judgment for our human frailties and sinfulness, and only prayer, penitence, and acts of justice/charity (one word in Hebrew, *zedaka!*) can avert the evil decree. And in the penitential prayers our ecological sins are placed alongside the catalogue of our other misdeeds for consideration.

In the main symbol of these days of awe, the *shofar* (ram's horn) and its shrill, primitive call to repentance, there is a hidden ecological note. We have in our past, in the ancient Temple service, animal sacrifice. But when it comes to the main symbol of atonement and the remembering of the sparing of Isaac on the mountaintop, it is not the whole animal, but an expendable portion that we use, the lack of which does no harm to the animal.

Ecology is involved in the basic reason for Rosh Hashanah. According to tradition this time of the year represents two beginnings: the creation of Adam and the creation of the cosmos. Human and cosmos share the same birthday; the one does not dominate the other. Somehow their destiny is linked, and part of the repentance/return (*teshvah*) has to do with our understanding that how we deal with our environment is directly involved in the case on our behalf in the heavenly court of judgment (Mandelbaum 1962, vol. 2, chap. 3).

Yom Kippur, the great fast day of intense introspection and self-evaluation, is a day on which we shut ourselves off from the world to prepare to return to that world a better and more caring person. It is on this day that we read the Book of Jonah as the prophetic reading that follows the afternoon Torah reading. As the day mounts to its climax, it is an ecological model that remains with us. For God reminds Jonah that the gourd, the cattle and the sheep, the fish of the sea, and the world of nature are a central concern. Jonah swallowed by the fish is indeed a somber warning. The reminder is well worth remembering: "You cared about the plant, for which you did not work and which you did not grow, which appeared overnight and perished overnight. And should I not care about Nineveh, that great city, in which there are more that a hundred and twenty thousand persons who do not know their right hand from their left hand, and much cattle as well?" (Jonah 4:10). Indeed, the gourd is the Bible's "spotted owl," held out as a warning to us to be ecologically responsible.

Just five days after Yom Kippur come the eight days of Tabernacles, of Sukkot, the ecological festival par excellence. The bridge from the God intoxication of the Day of Atonement back into the everyday world, that transition from holy to profane comes through the *Sukkah,* the frail booth in which we live for the eight days of the festival. We leave our houses of brick and mortar, our urban milieu, to live in a setting redolent with autumn's beauty, roofed with thatch, through which the stars can be seen. We carry the palm, willow, and myrtle tied together in one hand, and the citron in the other. We remember the wandering in the wilderness after Sinai, and just as we were brought closer to the divine in the days of awe, we are brought closer to nature around us and its needs.

When winter closes its grip on us, Hanukkah, a little minor festival—but how precious to children, and pregnant with ecological implications—comes to remind us of new years: a new year for trees. This festival is Tu B'Shvat, designated by its date, the fifteenth day of Sh'vat. In Israel and its Mediterranean setting, the almond trees began to bud in mid-January. So the trees have their festival. There is the talmudic story of a sage meeting an old man planting a tree. When asked why he is doing something the fruits of which he would never live to enjoy, he replied: "Just as my grandparents planted fruit trees from which I have eaten, so I plant them for my grandchildren" (Babylonian Talmud Ta'anit 23a).

And so it is a happy celebration, and in northern climates it comes in mid-winter as a welcome harbinger of spring.

How important this is seen in a statement made by Yohanan ben Zakkai about the Messiah: "If a man is planting a tree, and someone comes to him and says that the Messiah has come, let him finish planting the tree and then go to greet the Messiah!"

Passover comes with its message of exodus and freedom in the spring. The Seder table uses greens, and the reading of the Song of Songs highlights a sensitivity to spring and its hopes of rebirth, a rebirth as much for the environment as for us. Through the winter months, up to Passover, the core of the daily prayer includes a prayer for rain, reflecting the climate of the land of Israel where the winter rains were important to the spring and summer crops. From Passover through the summer, the prayer for rain shifts to a prayer for dew: always a concern for the right conditions for good crops.

Fifty days after Passover, the festival of freedom, comes Shavuot, the Feast of Weeks, also known as Pentecost. Something of ecology lies hidden in that fifty-day link. For the fifty days between these two festivals represents the fifty years of the Jubilee, the culmination of the seven sabbatical years with everything that this means for ecological healing. What applies to nature on the one hand applies to the link between freedom and Torah on the other.

Indeed, the three pilgrim festivals form a significant sequence in this sense. First come Passover and freedom: The newly freed slaves journey fifty days through the wilderness, come to Mount Sinai, and receive the Torah. Then comes Sukkot to remind them of the forty years of wandering in the desert until the entry to the Promised Land. And so we are bidden to count the days (*sefirat ha'omer*) from Passover to Shavuot, fifty of them, jubilee-preparing days, ecologically significant days.

On the thirty-third day we have Lag B'Omer—another great favorite of the ghetto-captive school children. This was the day they left their dingy classrooms and their dingy ghetto surroundings, and went out into the woods to be one again with nature. It was a happy, carefree time for them. It was also a special day for the mystics, for it was the day their sage and teacher, fleeing from the persecutions of the emperor Hadrian, emerged from the cave where he had hidden for seven years. It is on this day that we remember this sage, Simeon bar Yohai, who resisted the Romans, fled for his life, and hid in that cave where, tradition tells us, he was sustained by a fruit tree that sprouted at the mouth of the cave. When he finally emerged from the cave, he saw an old man (was it the man who planted the tree?) walking with myrtle branches in his hand. Asked where he was going he replied: "To greet the Sabbath." The myrtle branch and the Sabbath, Adam and cosmos, ecology and ethics! (B. Shabbat 33b).

When the celebration of Shavuot, the birthday of the Torah, comes, it is celebrated by all-night study of the Torah, by confirmation, by eating only vegetarian meals, and by decorating the synagogue with foliage. "Greens for Shavuot" was a slogan deep in the consciousness of many generations. A story with that title by the Yiddish writer Sholom-Aleichem burns brightly in my own memory of childhood (Sholom-Aleichem 1925, 15).

Moses de Leon, author of the Zohar, that great classic of kabbalism, in the thirteenth century spoke of the triad God, Torah, and Israel; and Franz Rosenzweig in the twentieth century spoke of creation, revelation, and redemption as the three basic principles of Judaism. Creation is at the beginning, and links humankind with cosmos. God's world and humankind's place in it: the human person as part of the cosmos on the one hand, and God's steward on the other, has profound implications for locating ecological concerns at the very heart and center of Judaism.

Conclusion

This chapter began with a blue box, a telephone call, and a later visit to the Valley of Hepher, now the breadbasket of Israel. It began with childhood memories that had to do with salvaging deserts, replanting forests, and making for a habitable world. The planting of a tree, the fragrance of the sukkah, the greens in the synagogue, the romp in the woods, the ram's horn, all woven in the fabric of memory, seem symbolically to have everything in common with what ecological issues are all about.

This two-pronged effort to enlist the rabbinic hermeneutic of biblical references and the cycle of the Jewish year is intended to suggest that Judaism is deeply connected with ecological values. It seems to come down to the tree. Do not destroy it. Make it a paradigm for everything that has to be done to protect the delicate ecological fabric that surrounds our world, so that we can save that world. As our "forefathers planted trees for us, may we plant trees for our grandchildren" (B. Ta'anith 23a).

REFERENCES

Soncino
1958 *The Talmud.* London: Soncino Press. 37 volumes.

Chekhov, Anton
1929 *The Works of Anton Chekhov.* New York.

Cohen, Jeremy
1989 *Be Fertile, and Increase, Fill the Earth and Master It.* Ithaca: Cornell University Press.

Freudenstein, Eric G.
1970 *"Ecology and the Jewish Tradition." Tradition* (Fall) 400–15.

Gordis, Robert
1929 *Judaic Ethics for a Lawless World.* New York: Jewish Theological Seminary.

Ha'am, Ahad
1921 *Al Parashat Derakhim* (The parting of the ways). 3:79.

Hamerton-Kelly, Robert G.
1992 *Sacred Violence.* Minneapolis: Fortress Press.

Harvard Divinity Bulletin
1991 "Theology for a Small Planet." (Fall).

Heschel, Abraham J.
1955 *God In Search of Man.* New York: Farrar, Strauss and Cudahy.

Mandelbaum, Bernard
1962 *Pesikta d'Rav Kahana.* New York: Jewish Theological Seminary.

Pelkowitz, Ralph
1976 *Danger and Opportunity.* New York: Shengold.

Perelmuter, Hayim Goren
1989 *Siblings.* Mahwah, N.J.: Paulist Press.

Sholom-Aleichem
1925 *Collected Works* [in Yiddish]. New York.

Toynbee, Arnold
1971 International Journal of Environmental Studies.

White, Lynn
1967 "Historical Roots of the Ecological Crisis." *Science* 155:1203–7.

Contributors

Dianne Bergant, C.S.A., is professor of Old Testament studies at Catholic Theological Union in Chicago. She is a member of the North American Conference on Religion and Ecology and has written several articles on issues of Bible and ecology. Her major publications include *The World is a Prayerful Place, Collegeville Bible Commentary* (OT editor), and "The Wisdom Books," *The Catholic Study Bible* (New York, 1990).

Edward Foley, Capuchin, is associate professor of Liturgy and Music at Catholic Theological Union. Editor of the *American Essays in Liturgy* series, his recent publications include *From Age to Age* and *First Ordinary of the Royal Abbey of St.-Denis in France* (Paris, Bibliotheque Mazarine 526).

Richard N. Fragomeni, a presbyter of the diocese of Albany, New York, is assistant professor of liturgy and homiletics at Catholic Theological Union. An internationally known speaker on the catechumenate, along with his many published articles he has recently co-edited *A Promise of Presence: Studies in Honor of David N. Power.*

Kathleen Hughes, a religious of the Sacred Heart and professor of liturgy, is vice-president and academic dean of Catholic Theological Union. Among her publications, the most recent are *The Monk's Tale: A Biography of Godfrey Diekmann, O.S.B.* and *Living No Longer for Ourselves: Liturgy and Justice in the 90's,* the latter co-edited with Mark Francis.

John M. Lozano, a Claretian priest, is professor of spiritual theology at Catholic Theological Union, and has written extensively in his fields of interest. Among his more recent books are *Prayer Even When the Door Seems Closed* and *Life as Parable: Reinterpreting Religious Life.*

Thomas McGonigle, O.P., was formerly associate professor of Church history and Spirituality and Vice-President and Academic Dean of Catholic Theological Union. A noted medievalist and historian of the Reformation, he serves as vice-president of academic affairs at Providence College in Rhode Island.

Thomas A. Nairn, O.F.M., is associate professor of ethics at Catholic Theological Union. He serves as medical ethics consultant at the Alexian Brothers Medical Center, Elk Grove Village, Illinois. His scholarly articles have appeared in *New*

Theology Review, Journal of the Catholic Campus Ministry Association, and *Process Studies.*

Gilbert Ostdiek, O.F.M., is professor of liturgy at Catholic Theological Union and co-director of the Institute for Liturgical Consultants. He serves on the advisory committee of the International Commission on English in the Liturgy (ICEL) and chairs its subcommittee on translation and revision of texts. He is past-president of the North American Academy of Liturgy (NAAL). His publications include *Catechesis for Liturgy: A Program for Parish Involvement.*

John T. Pawlikowski, a Servite priest, is professor of social ethics at Catholic Theological Union. He is senior editor for *New Theology Review* and serves as a member of the editorial boards of other journals, which include *SHOFAR: An Interdisciplinary Journal of Jewish Studies, Journal of Ecumenical Studies,* and the *Journal of Holocaust and Genocide Studies.* He is a prolific writer in the area of Jewish/Christian relations and is a presidential appointee to the U.S. Holocaust Memorial Council. He is the author of ten book and numerous articles.

Hayim Goren Perelmuter, Rabbi Emeritus of KAM Isaiah Israel Congregation, is currently Chautauqua Professor of Jewish Studies at the Catholic Theological Union, on which faculty he has served since 1968. He has written extensively, and has translated from Hebrew, Yiddish, and German. Among his noted books are *Siblings: Rabbinic Judaism and Early Christianity at Their Beginnings* and *David Darshan's Song of the Steps and In Defence of Preachers.*

Barbara E. Reid, O.P., is a Grand Rapids Dominican. She is assistant Professor of New Testament studies at Catholic Theological Union. Her scholarly interests center on Luke-Acts, Paul, and feminist interpretation of Scripture. Her recent articles can be found in *Biblical Research, The Bible Today,* and *New Theology Review.*

Paul J. Wadell, C.P., is associate professor of ethics at Catholic Theological Union. He is author of *Friendship and the Moral Life, The Primacy of Love: An Introduction to the Ethics of Thomas Aquinas,* and *Friends of God: Virtues and Gifts in Aquinas,* as well as several articles in Christian ethics and spirituality.